GREEN PRODUCTIVITY IN SMALL AND MEDIUM ENTERPRISES

By the Same Author

Small Scale Industries (In 2 Volumes)

About the Editor

Dr. M. Soundarapandian is Professor of Rural Industries and Management and Director, Centre for Entrepreneurship Development, Gandhigram Rural Institute (Deemed University), Gandhigram. Apart from 22 years of teaching and extension responsibilities at Gandhigram Rural Institute, he has been engaged in reserch activities.

He has published many research articles in various journals at national and international level and eighteen books, namely, Rural Industries in India, Women Entrepreneurship, Tribal Development, Industrial Resurgence in Rural India, Literacy Campaign in India, Rural Entrepreneurship, Research Projects in Rural Industries and Management, Rural Industries Under Liberalisation, Rural Industrialisation, Small Scale Industries, Economic Reforms & Public Distribution System, Management Model for Self-employment, Economic Reforms and Employment, Environment Degradation and Rural Poverty, Textile Industry under Globalisation, Rural Labour Market etc.

Seven scholars were awarded Ph.D. Degree under his guidance and supervision. Apart from providing research guidance to a number of Ph.D. and M.Phil Scholars, he has directed and completed eleven minor Research Projects and four major Research Projects that were sponsored by University Grants Commission, Indian Council of Social Science Research, Bharat Gyan Vijayan Samiti, Ministry of Rural Development, New Delhi and State Resource Centre for Non-formal Education, Chennai. The Young Economist Award by the Association of Economists of Tamil Nadu and Bharat Jothi Award by the India International Friendship Society have been awarded to him.

GREEN PRODUCTIVITY IN SMALL AND MEDIUM ENTERPRISES

Volume–I

(Industry and Agriculture)

Edited by

M. Soundarapandian
Professor of Rural Industries and Management,
Director,
Centre for Entrepreneurship Development (CED)
Gandhigram Rural Institute
(Deemed University)
Gandhigram–624302

CONCEPT PUBLISHING COMPANY, NEW DELHI-110059

Cataloging in Publication Data—DK
Courtesy: D.K. Agencies (P) Ltd. <docinfo@dkagencies.com>

Green productivity in small and medium enterprises / edited by M. Soundarapandian.
2 v. cm.
Papers presented at a national seminar held at Gandhigram; with special reference to India.
Includes bibliographical references.
Includes indexes.
Contents: v. 1. Industry and agriculture — v. 2. Environment, health and waste management.
ISBN-13: 9788180693700 (set) ISBN-10: 8180693708 (set)
ISBN-13: 9788180693717 (v. 1) ISBN-10: 8180693716 (v. 1)
ISBN-13: 9788180693724 (v. 2) ISBN-10: 8180693724 (v. 2)

1. Small business—Environmental aspects—India—Congresses. 2. Agriculture—Environmental aspects—India—Congresses. 3. Environmental health—India—Congresses. 4. Refuse and refuse disposal—India—Congresses. I. Soundarapandian, M. (Mookkiah), 1960-

DDC 338.642 22

ISBN-13 : 978-81-8069-370-0 (Set) ISBN-10 : 81-8069-370-8 (Set)
ISBN-13 : 978-81-8069-371-7 (Vol-I) ISBN-10 : 81-8069-371-6 (Vol-I)
ISBN-13 : 978-81-8069-372-4 (Vol-II) ISBN-10 : 81-8069-372-4(Vol-II)

First Published 2007

Published and Printed by
Ashok Kumar Mittal
Concept Publishing Company
A/15-16, Commercial Block, Mohan Garden
New Delhi-110059 (India)

Phones : 25351460, 25351794
Fax : +91-11-25357103
Email : publishing@conceptpub.com

PREFACE

Green productivity is not a new concept but has emerged gradually. Because it concentrates upon the preventive approach, it assumes the utmost importance as far as Small-scale industries are concerned. Green productivity can be explained as adoption and implementation of environmentally sound processes and procedures to optimize the utilization of resources to enhance production. In other words, this concept focuses upon enhancing, producing without rating adverse impacts on the environment. Seen from a different perspective, improvement of the environment should have beneficial rather than deleterious effects on the quantity and quality of production. Thus green productivity stresses optimum utilization of space, manpower, energy and raw materials, i.e., all the factors of production.

In view of strict implementation of existing environment regulations and more regulations to be promulgated in the future, two things have become clear. First, it will be difficult for small and medium enterprises to continue to operate. For example, in India many small-scale units have been forced to close or change location for environmental reasons. Second, if SMEs invest in waste (end of Pipe) treatment, their production costs may not remain competitive. Now SMEs are finding themselves, in a dilemma, and the wayout appears to be the green productivity approach because it results in a clearer environment and higher production at less cost.

Green productivity can resolve a number of major problems of SMEs pertaining to the environment. It can enable them to sustain today's cutthroat competition and survive against giant multinational companies. On the other hand, it will help industries to create a positive image as an environmentally responsible company in the market, while at the same time helping the company to confirm to regulatory requirements. Once the concept of green productivity is understood, the question arises of how to attain green productivity and make it a perpetual movement.

Based on the above background, a national seminar was organized with the assistance of AICTE, New Delhi.

Selected papers presented in the seminar are edited as a book on "Green Productivity in Small and Medium Enterprises". The book has been divided into four parts, viz., Industry, Agriculture, Waste Management and Environment and Health. The book is very useful to the young researchers, academicians and the policy-makers.

I am thankful to Dr. T. Karanakaran, Vice-Chancellor and Dr. N. Meenakshisundaram, Professor and Head, Department of Rural Industries and Management for their encouragement in completing the work. It is my pleasure to thank my wife Mrs. S. Indumathi and my children S.Diwakaran and S.Vidya for their kind cooperation. Finally, I am grateful to the Concept Publishing Company, New Delhi for publishing the book.

Gandhigram **Dr. M. Soundarapandian**

CONTENTS

LIST OF TABLES AND FIGURES

Figures

LIST OF THE CONTRIBUTORS

A. Abdulraheem,
Dept. of Economics,
The New college
Chennai

A. Baradeswaran
Dept. of Mechnanical Engineering,
Arunari Engineering College,
Tiruvannamalai.

A. Chandraprabha
Dept. of Rural Industries and Management
Gandhigram Rural Institute
Gandhigram.

A. Meenakshisundarajan,
Dept. of Economics
S.J. Hindu College
Nagercoil.

A. Saravana Naidu
Dept. of RIM
Gandhigram Rural Institute
Gandhigram.

A. Thaha Sahad
Dept. of Economics,
The New College
Chennai

C.A. Sham Shankar
Dept. of Economics
S.J. Hindu College
Nagercoil.

C. Parvathi
Dept. of Economics,
Avinashilingam University,
Coimbatore

C. Sekar
Dept, of Agricultural Economics
Tamil Nadu Agricultural University,
Coimbatore.

Dr. D. Solomon Raj
Reader in Economics,
Pope's College,
Sowyerpa maro-62825

Er. E.T. Narasimhan
Dept. of RIM
Gandhigram Rural Institute
Gandhigram

G. Anoop
Dept. of Agricultural Economics
Tamil Nadu Agricultural University,
Coimbatore.

G. Saravanan
Dept. of RIM,
Gandhigram Rural Institute,
Gandhigram.

G. Somasundaram
Dept, of Agricultural Economics
Tamil Nadu Agricultural University,
Coimbatore.

I. Jayalakshmi
Lecturer,
Dept. of Textiles & Clothing,
Avinashilingam University,
Coimbatore

I. Sundar
Senior Lecturer in Business Administration
D.D.E., Annamalai University,
Annamalai Nagar.

K. Balasubramani
Dept. of RIM,
Gandhigram Rural University,
Gandhigram.

L. Radhakrishnan
Dept. of RIM
Gandhigram Rural Institute
Gandhigram.

M. Anjugam
Dept. of Agricultural Economics
Tamil Nadu Agricultural University,
Coimbatore.

M. Sivakumar
Dept. of Management Studies,
PSNA College of Engineering & Technology, Dindigul

M. Soundarapandian
Dept. of Rural Industries and Management
Gandhigram Rural Institute
Gandhigram.

N. Manjula
Dept. of Mangement Studies,
KLN College of Engineering
Pottapalayam

N. Murugan
Dept. of Management Studies,
PSNA College of Engineering & Technology, Dindigul

N. Prabha
Dept. of RIM
Gandhigram Rural Institute
Gandhigram

P. Balaji
Tamil Nadu Agricultural University
Coimbatore.

P. Namboothiripad,
Tamil Nadu Agricultural University
Coimbatore.

P. R. Wadje
DKTE Society's Textile & Engineering Institute,
Ichalkaramji

R. Seerangarajan
Reader, Dept. of RIM
Gandhigram Rural Institute
Gandhigram

S. Amsamani
Reader, Dept. of Textile & Clothing
Avinashilingam University
Coimbatore.

S. Gurunathan,
Tamil Nadu Agricultural University
Coimbatore.

Samwel Kakuko Lopoyetum
Post Doctoral Research Fellow
Department of Cooperation
Gandhigram Rural Institute
Gandhigram.

S.N. Sukumar
Reader in Economics
RKM Vivakanda Collage,
Channai.

S.N.R. Srikumar
Faculty ot Management Studies,
KLN College of Engineering,
Pottapalayam

S. Radhakrishnan
Dept. of RIM,
Gandhigram Rural University,
Gandhigram.

S. Subbiah
Senior Lecturer in Business Administration
D.D.E., Annamalai University,
Annamalai Nagar.

S. Syed Ahamed
Dept. of Mechnanical Engineering,
Arunari Engineering College,
Tiruvannamalai.

T. Karthikeyan
Dept. Managment Studies,
Annamalai University
Annamalai Nager.

V. Balakrishnan
Dept. Managment Studies,
Annamalai University
Annamalai Nager.

V.B. Upadhye
DKTE Society's Textile &
Engineering Institute,
Ichalkaramji

V. Nachimuthu
Dept. of Economics,
Srivasari College
Erode.

Mrs. V. Rajam
Dept. of Economics,
Avinashilingam University,
Coimbatore.

V.S. Ganesamurthy,
Dept. of Economics,
Srivasari College
Erode.

1

APPLICATION OF GREEN PRODUCTIVITY CONCEPTS IN THE MANAGEMENT OF INDUSTRIAL ESTATES

I. Sundar

Introduction

Environmental protection through pollution prevention and the meeting of environmental standards by waste treatment of the effluents in the various industries have not worked in many countries. The problem of industrial environmental pollution is particularly serious in developing countries where the enforcement of environmental regulations is not strictly enforced. Environment protection is seen by industries as only an added cost, which reduces competitiveness and profits of the enterprises that strictly follow such environmental regulations. This is particularly true of small and medium scale industries that usually do not have the financial and human resources to manage environmental problems.

Environment protection needs to be accompanied by productivity and quality improvements if it is to be more widely accepted and practiced by the industries. Even when industries can meet environmental quality standards, environmental protection alone is also seen as insufficient for sustainable development, though it is a component of it. Environmental protection does not concern itself with broader issues of natural resource use, biodiversity and the ecological impacts of pollution. The sustainability of the environment is at risk through the over-exploitation of natural resources and ecological disruptions through pollution and ecosystem destruction that usually result from development activities.

By keeping these facts in mind this chapter analyses the implications of Green Productivity concepts in the management of industrial estates. Green Productivity is not seen as an end in itself, but must bring social and economic benefits to the people. The improved environment that results from the practice of Green Productivity will translate into cleaner and healthier living conditions and an improved quality of life, within the workplace as well as in the surrounding areas. In industries, the direct benefits of increased profitability accrue to the owners, which may result in some financial benefits and better working conditions for the workers depending on the decision of management.

CONCEPTS OF GREEN PRODUCTIVITY

There are a number of concepts that guide the practice of Green Productivity. It is made up of both environmental and productivity concepts. The ecological sustainability of the environment is the basis of economic and social sustainability. Ecological sustainability, is made up of three key elements, as follows. Green Productivity aims at conservation of natural resources would mean that there would be the efficient use of resources. This through efficient usage, recycling and reuse wherever possible. An underlying assumption for sustainable use is that the rate of use of renewable resources should not be greater than the rate of renewal and for non-renewable resources, the rate of use should not be greater than the rate that substitutes can be found.

Green Productivity plays a significant role in protection of life support system. It could be noted that ecological processes are the basis of balance within the environment that provides us with clean air and water and other ecological services making life possible. Pollution of the environment disrupts the ecological balance that exists in nature. This can result in adverse human health impacts such as the accumulation of toxic chemicals in the ecological food chain. The pollution levels that enter the environment should not exceed the absorptive and regenerative capacity of the environment.

The conservation of biological diversity is one of the aspects of Green Productivity. It is significant to observe that our economic activities and in particular, the industrial production processes, should not erode or destroy biological diversity. Besides its intrinsic values, biological diversity has the great potential of contributing to our future wellbeing in terms of the potential useful products that may be needed in future.

Eco-efficiency is another important constituent of green production. Eco-effciency is now an objective of many industries. Various tools and technique have been used to achieve this. Eco-efficiency, with its emphasis on producing more with less and reducing impacts on the environment is a key concept in Green Productivity.

Productivity Concepts

Continuous Productivity Improvement

The concept of continuous incremental improvement has been one of the key concepts in the productivity improvement movement over the last 40 years so in many countries. The Deming cycle is the main means of achieving the continuous improvement in the productive process. The step-by-step problem solving framework or cycle of the Green Productivity methodology is a variation of the Deming cycle. Through the practice of this cycle, continuous improvement is achieved.

Further Quality Improvement should be considered carefully. Quality improvement is the other key concept of the productivity movement and many tools and techniques had been developed to achieve that Green Productivity makes use of many of these tools and techniques, in conjunction with the step-by-step methodology.

Integration of Environmental and Productivity Objectives

It could be achieved in two ways—conceptual integration and methodological integration with the acceptance that productivity and environmental objectives are of equal importance, the planning for productivity improvement would always be guided by environmental considerations. In the way that productivity and quality are now synonymous, in future, environment and productivity would be synonymous. Methodological integration takes place in the way the problem solving process is carried out and in the choice of the tools, techniques and technologies that would be the most appropriate. This would include both productivity and environmental protection tools and techniques.

PRINCIPLES OF GREEN PRODUCTIVITY

Accountability is one of the aspects of industrial development and its consequence on environment. The principle that industries can be held responsible for the consequences of their activity and products is becoming increasingly accepted and in some countries, such as in Europe and Japan, legislative measures are being introduced to ensure compliance. In some of these countries, there is the compulsory take back of the products from the consumers after the end of the useful life of the product. Green Productivity attempts to incorporate this principle in its practice. Polluter Pay Principle is the next aspect of Green Productivity.

This principle is being put into practice in more and more countries as the cost of making good the environmental damage is charged to the responsible party. Green Productivity practice minimizes the pollution in the production process.

Precautionary principle is one of the constituents of green production. The precautionary principle, a guiding principle of Green Productivity, dictates that one adopt a cautious approach even when the evidence is not irrefutable, but the effects are long term and difficult to rectify and reverse. The reduction or avoidance of the use of hazardous material is a common practice under Green Productivity when the disposal and treatment of the chemicals is uncertain. The issue of global climate change is one example where precautionary principle should be applied. Global warming is caused by the large-scale introduction of greenhouse gases from internal combustion engines and factories. There had been some scientific uncertainty on the extent of the long-term impacts of the increase of greenhouse gases in the atmosphere, though there has been growing scientific consensus that the impacts are serious, real and almost irreversible in the human time scale.

Principle of Inter-generational and Intra-generational Equity forms a sustainable focus of Green Productivity concept. The issue of the equitable sharing of benefits between generations (or inter-generation equity) was highlighted in Agenda 21 and encapsulated in the definition of sustainable development. The present generation does not have the right to use up all the resources and pollute the environment to the extent that future generations would not be able to satisfy their livelihood. The inequitable use of resources in the present time amongst nations and within nations (intra-generational equity) is an issue that is part of Agenda 21. This issue is partially being addressed in Green Productivity where the emphasis is on improving the quality of life of the people.

Productivity Principles

There are four productivity principles, viz., profitability, competitive advantage, people centred orientation and multi-stakeholder participation. Regarding profitability principle economic viability of the enterprise practicing Green Productivity is one of the underlying requirements of Green Productivity. The increase efficiency in resource use and production processes that usually accompany the application of Green Productivity is translated into savings in cost and hence in profitability. The profitability of Green Productivity is a powerful incentive for the practice of Green Productivity by industries.

The second one is competitive advantage, which brings positive business outcome. It is observed that the practice of Green Productivity results in several competitive advantages to the enterprises in terms of greater marketing opportunities. An added benefit is the compliance to legal environmental, requirements.

In the case of people Centred Orientation the sharing of benefits among the people who contributed to the success of the project would contribute to social stability and long-term sustainability is of vital importance. This is more a value judgment than an operating principle as it is still possible to adopt Green Productivity without the sharing of economic benefits. However, in a number of Green Productivity demonstration projects, there had been significant sharing of the benefits arising from Green Productivity implementation.

Further multi-stakeholder participation is one of the production principles of Green Productivity. It is a principle that underlies the practice of Green Productivity to solve many of the problems encountered during and production and distribution of goods and services. Even from a practical point of view in terms of the effectiveness of a project, multi-stakeholder participation often ensures the success of the Green Productivity project. Many projects require the cooperative effort of all the major stakeholders.

Green Productivity and Industrial Estate Management

As the industrial estate is made up of many individual factories, the practice of Green Productivity by all the individual factories would achieve the objectives of productivity improvement and environmental protection. The cumulative effect of individual factories applying Green Productivity would reduce the overall environmental problems of the industrial estates. The problem would be to convince all the factories to practice Green Productivity at the factory level especially when there is no problem of crisis proportions, such as threat to shut down the factory because of a failure to persistently meet environmental regulations and standards.

In practice, not all factories will be able to solve all their environmental problems even if they practice Green Productivity. This may be due to the cost of implementing the Green Productivity measures, old machinery, and outdated processes, which have limited financial resources and expertise. The small size of some of the factories may preclude certain solutions such as resource recovery when the amount from one individual may be too little to warrant investment on a large scale to make it economical. The role of management of industrial estates may be to get the factories to cooperate in the treatment of the waste residues, such as the pooling of resources for waste treatment, for resource recovery and for recycling. The Green Productivity approach would be applicable once a problem has been defined and agreed upon.

Industrial Park Scale

An industrial park can also be treated as a single entity and the Green Productivity concepts and methodology can be applied to the industrial estate. Green Productivity is closely related to the Environment Management System (EMS). The environmental policy of the industrial estate can be established and the implementation of the EMS will be facilitated with the implementation of Green Productivity. The sharing of common recovery and treatment facilities for waste management would be feasible and desirable at the level of industrial estates, especially for similar industries. The management of the industrial estate could be extended to by-product exchange between industries and the sharing of various facilities. When this level of cooperation is achieved, it would approach the practice of some of the eco-industrial estates.

The potential and opportunity of managing an industrial park on an ecologically sound basis would be highest for new industrial parks when the planning and development incorporates eco-industrial park principles at the very beginning. However, even in an established industrial park, there would be opportunities for changes because all industries require renovation and redevelopment constantly in an increasing competitive environment. With environment being an increasingly important aspect of trade and industry, both locally and globally, the incorporation of environmental and sustainability concerns in industrial estates will become correspondingly important.

Problems in application of Green Productivity Industrial Estate Management

Some of the methodological problems are observed in application of Green Productivity in management of industrial estate. It is not always clear what the equivalent function or role would be between a factory and an industrial estate, as the structure and objectives of the two different levels of organization would differ. The methodology of Green Productivity work, therefore, need to be adapted for the industrial estate. The kind of methodological adjustment that has to be made is similar to that on the application of the EMS and ISO-14000 from factories to large institutions and municipal councils. A similar adaptation took place when the methodology was applied to problems of the community, rather than a factory. While in theory Green Productivity can be applied at the level of the industrial park, this has not been tried out so far. The adaptation of methodology would need to look into the problem hierarchy level and scale as the industrial estate is at a different organizational level compared to a factory.

It is interesting to observe some more discussions of Green Productivity. From Environmental Protection to Sustainable Development, the trend in environmental policy is the move beyond environmental protection to sustainability. Most countries have accepted the concept of sustainable development. As they attempt to make the transition towards sustainable development, the framework of environmental protection alone is seen as too limited and cannot be the sole basis for sustainable development. The conceptual basis for the development of Green Productivity applications should, therefore, reflect sustainability concerns and not focus too narrowly on environment alone. The social and economic dimensions of development are as critical as the environmental dimension in the sustainable development. Also in sustainable development, the process is as important as the results.

Systems Approach to Environment

The system approach to the environment is an approach whereby the natural ecosystems are seen as models for the development of industrial systems and other man made systems. Mimicking natural systems will be seen as possible ways of achieving eco-efficiency and sustainability, as the natural ecosystems had evolved over long periods of time to reach some sort of dynamic balance. The various ecological processes that link the components of the ecosystem will be look upon as examples of how to achieve ecological balance and sustainability. As within the natural ecosystems, there is no waste as such as the output of one process is the input of another. Industrial ecology is one of the approaches seen to be adopting a systems approach to problem of the industry. A manifestation of this approach is the concept of industrial parks where there is an attempt to achieve industrial symbiosis between factories. The terminology used is ecological and biological, rather than engineering as in the past.

The development of the Green Productivity approach has had a strong engineering bias as the initial applications focused on the environmental problems of the manufacturing industries. However, this is changing as the applications of Green Productivity have broadened to include the services industry such as tourism and eco-tourism where management and natural resources conservation issues are more important. Eco-tourism in particular requires the understanding of structure and functions of the natural ecosystems.

There is a need to analyse the Inter-connectedness of human activities in relation to management of industrial estates. It is increasingly recognized that human activities are inter-connected and that one needs to take a more integrative and holistic approach to solve problems. The consequences of action in one aspect of human activity can have often unanticipated impacts on other activities. This would mean dealing with the problems of making the transition to sustainable development. There would need to be more people from different disciplines involved in the solution of the problems. This is particularly evident in some of the issues in community development where in addition to persons with backgrounds in environmental sciences and engineering, there would be a need for persons in agriculture, ecology, economics, sociology, management, planning, political science, law and other fields. The broadening of the scope of Green Productivity to include areas like eco-industrial park planning and management, the development of rural and urban communities, and the greening of municipal and city councils would mean that the conceptual and methodological frameworks would need to be adapted and developed further.

Analyses of long-term perspective are essential to study the sustainability. With sustainability as a major concern, the perspective of development will have to take a longer perspective and not only in the time frame of 5 year development plans, as sustainability is measured in terms of generations rather than years.

At the present moment, it is difficult to monitor the impacts of Green Productivity to beyond the duration of the project cycle. Nevertheless, it would be possible to devise a regular monitoring programme to evaluate the Green Productivity programmes after a number of years.

The scope of Green Productivity is constantly broadening to include more economic sectors and a wider range of human activities, such as community development and management of industrial estates. The sustainability of industries would become more critical as more and more of the developing countries embark on ambitious industrialization programmes. With the economic expansion projected to increase 5 to 10 times in Asia over the next 30 to 40 years, the tasks to ensure the sustainability of industry and of development takes on an added urgency. The issue of sustainability is a challenge that Green Productivity is attempting to meet.

GREEN PRODUCTIVITY AND POLLUTION PREVENTION

Source Reduction

The Source Reduction technique includes practices that reduce the amount or toxicity of waste, hazardous substance, pollutant, or contaminant entering a waste stream or otherwise released to the environment prior to external recycling, treatment or disposal. Source reduction measures generally fall into the following categories.

Input Material Changes

Hazardous substances can be introduced into a process through the input materials used to manufacture a product. These hazardous materials can be present in both primary (raw) materials used to manufacture a product (e.g. hydrocarbons used to make plastics) and secondary materials. The latter is not a component but is being consumed in the process (e.g. solvents used in cleaning process equipment). Input material changes fall into two major categories:

> *Material Substitution* is the substitution of existing material inputs to a process by materials that are environmentally friendlier, without adversely affecting the product quality.
> *Material Purification* is when raw materials that are of higher purity are used to reduce the quantity of wastes generated.

Process Change

This method is concerned with how the product is made. This would include *Technology Change* and *Improved Operating Procedure*. Such changes reduce workers' exposure to the pollutants during the manufacturing process. Process changes may be implemented more quickly than product changes. Typically, improved operating procedures can also be implemented more quickly and inexpensively than input material and technology changes.

> *Technology Change* involves modifying the technology used to produce a product or service. It is one of the most effective methods of preventing pollution generation. This method is less common among companies as it generally involves greater personnel and capital investment, and the result often takes a longer time to realize.
> *Improved Operating Procedures* can be implemented in all areas—production; maintenance; raw material, product, and waste handling; and storage. Since good operating procedures can often be implemented at a low cost, they usually have a higher return on investment. Procedural aspects of a manufacturing operation include the management, organizational, and personnel functions of production. Efforts in improving operating procedures could be Scheduling Improvements, Material Handling and Storage, Spill and Leak Prevention, of Personnel Practices.

Waste Stream Segregation

Waste Stream Segregation is one of the aspects of pollution control waste. Stream Segregation refers to the separation of waste streams according to points of generation, composition, volume or media. Hazardous waste that is normally hauled to off-site disposal facilities is often a combination of two or more waste stream types. Segregation at the source can reduce the quantity of disposal of hazardous wastes. When a non-hazardous waste is mixed with a listed hazardous waste, the entire mixture is classified as hazardous; not allowing hazardous and non-hazardous wastes to mix reduces the amount of hazardous waste that requires disposal and yields substantial savings. Some benefits of Waste Segregation include: ease in endow-pipe treatment of a non-compatible pollutant stream and increased possibility of recycling/reusing a waste stream.

Energy Conservation

Production facilities consume energy basically in two different forms : electricity and process heat. Combustion of fossil fuels in primary heat sources such as boilers or fired heaters provides a major source of heat input to industrial processes. In the techniques of *Energy Conversion*, the focus is on the energy conversion efficiency of the units such as industrial boilers, atomic power reactors, etc. that convert fuel to steam, or fuel to electricity. Alternately, one can also focus on *Energy Transfer* where the energy transfer efficiency of the energy conducts and steam piping can contribute significantly to reduce losses during operation. We can also conserve energy through reducing *Energy Utilization* by improving the individual equipment efficiency in terms of unit product output per energy utilized. In situation where energy utilization is already in an optimum state, *Energy Recovery* techniques can be used. Heat exchangers are deployed to draw the excess heat in the effluent and the recovered heat is subsequently used back in the process.

Recycle, Reuse and Recovery

Recycle, Reuse and Recovery is considered as the second most preferred methods after source reduction techniques in the environmental management hierarchy. This methodology is advantageous because of the following reasons: conservation of natural resources, elimination of waste management options like treatment or land disposal and reduction in raw materials needed in the processes. However, Recycle, Reuse and Recovery can incur somewhat increased risk and liability due to additional handling and management of materials. Thus, this may affect the quality of the final product.

Recycle is a technique where used materials, energy or leftovers in some cases are recycled within the production facility. For example, scrap metal or old parts can be melted down and cast into new material for processing. Heat generated in some processes can be recycled through waste heat exchangers to heat other parts of the process.

Reuse is technique where the waste materials generated in a manufacturing process are subsequently reused in the original process with or without treatment to remove impurities. For example, material containers can be reconditioned and re-used with minimal efforts. If waste material cannot be directly reused in the original process, because of potential contamination, it may be treated to remove contaminants. For example, organic solvents used in parts leaning and pharmaceutical manufacturing processes are often collected, distilled, and reused in the original process.

Recovery is a technique wherc valuable resources can be reclaimed from wastes in the form of raw materials, by-products or products. Recovery normally is the preceding activity to recycle or reuse. *On-site Recovery* is practiced by industries to recover resources that can be reused in the process or sold. *On-site Recycling* is employed when the resource can be easily segregated; when it has a substantial reuse value to the industry or an external market and when there are no risks involved to the industry in the recovery process. When On-site reuse options are not feasible, then the next step is to consider *Off Site Recovery and Recycling* option to save treatment/disposal costs. The recycled material could be either returned to the generator for reuse, or sold for use at other facilities.

End-of-Pipe Treatment Technologies

End-of-Pipe treatment aims to reduce waste generation to the point where treatment and subsequent disposal in the environment can be avoided. However, there are still some processes in which waste generation is inevitable. This waste, which cannot be eliminated, reduced, recycled or reused, must be treated and disposed within all applicable environmental regulations. Appropriate treatments that are in-line with environmental regulations do not include transferring pollutants to other environmental media or dilution. There are a number of end-of-pipe pollution control measures that are applicable to treat and dispose the pollutants generated.

Waste Treatment Processes are practiced widely in the industries. These processes involve changing the form or composition of a waste stream through control reactions to reduce or eliminate the amount of pollutants produced. Examples are detoxification, incineration, decomposition, stabilization, and solidification, or encapsulation. Many control practices involve collecting the pollutants and moving them from one environmental medium to another. An example of the control practices is scrubbing. It is used to remove sulfur compounds from flumes of combustion process.

Volume Reduction refers to reducing the volume of the waste such that it can be easily managed, handled and disposed of in an economic manner. These approaches usually result in high concentration of the hazardous or toxic constituents. For example, pressure filtration and drying of a heavy metal containing sludge prior to disposal decreases the sludge water content and waste volume, thereby reducing the cost of disposal.

Designing Environmentally Compatible Products

Environmentally compatible products minimize the adverse impacts on the environment resulting from their manufacture, use and disposal. The extent of the environmental impact of a product is determined during its design phase. By taking environmental considerations into account during product planning, design and development, a company can minimize the negative impact of its products on the environment. Modern techniques like *Design for Environment (DFE)* and *Life Cycle Assessment (LCA)* can be used in the process of designing environmentally compatible products.

Design changes made to prevent pollution should be implemented in a manner that the quality or function of the product is not adversely affected. Design for the environment can be achieved by the people directly involved or people that are within the framework of company policy. The resulting product ranges would include Product Conservation and Product Design Changes. Lower production costs through efficient resource utilization. Improve product performance Reduce environmental impact.

Product Conservation refers to the way in which an end product is used. For example, better maintenance of process equipment and components by industry can decrease the frequency of equipment component replacement, which in turn reduces the waste generated by the used component. A manufacturer can also alter a product to minimize the waste resulting from the product's end-use. For example, a manufacturer of lubricants could develop a product that lasts longer than conventional lubricants, thereby reducing the amount of waste lubricant generated.

Product Design Changes involve manufacturing a product with a lower composition of hazardous substances or changing the products' composition so that no hazardous substances are involved. For example, a manufacturer could use an active ingredient in a formula with a non-hazardous solvent rather than a chlorinated solvent. Other examples include using mineral oil in electrical transformers instead of polychlorinated biphenyl (PCB) liquids or using organic pigments in paints rather than heavy metal pigments.

Wastes of Production

We often think of waste as physical items like materials, water, electricity, etc. However, in the context of Productivity, waste is defined as : Any form of work, which does not add value to the final output.

Waiting is one form of waste that we encounter everywhere. For example, workers waiting for a machine may not be functioning and delay in arrival of materials. The causes of such waste are often lack of planning, lack of proper training, lack of control and lack of discipline.

Transport waste refers to moving things from one place to another unnecessarily. Such actions do not add value to the product. Hence, it should be eliminated or reduced as far as possible. There are two aspects to be considered: improving the floor layout and improving the method of transport by using more advanced material handling equipment.

Processing waste is inherent in the production process or design. For example: an electronic type-writer has fewer parts and processes than a mechanical type-writer, replacing a metal dust-bin with a plastic dust-bin can reduce several steps in the production process and using preprinted forms can save a lot of time during documentation.

Inventory waste occurs when the inventory turnover is low. This will consume valuable factory space and financial resources. Inventory wastes also occur when inventory has become obsolete.

Motion waste refers to the unnecessary movement of the body that does not add value to the work we do. Motion study is one aspect of Industrial Engineering that assists us to reduce waste in motion. We can reduce motion waste by improving the workplace layout, practicing good housekeeping and workplace organization, introducing jigs and fixtures and low cost automation.

Defect waste is waste caused by producing bad quality products or poor service. Time is wasted when reworking the bad products or addressing customer complaints. Additional manpower and space is often required to handle the defective items.

Overproduction waste refers to producing more than what is actually needed. The unused products may have to be discarded when not required at a later stage. This waste will ultimately increase the unit cost of production. Poor planning, poor forecasting and producing too early are common causes of overproduction waste.

Good Housekeeping Practices

Good housekeeping is not limited to keeping the work place clean and eliminating leaks and spills. It also includes improving operational procedures and systems to make them more resource efficient. Some examples of Good Housekeeping Practices are preparation of recipes in the right quantity to avoid surplus, efficient handling of materials, optimum storage procedures to avoid losses and material degradation during storage. In small-scale industries, good Housekeeping Practices alone could lead to a reduction in waste generation of up to 20-25 per cent. It is a set of workplace management techniques that focuses on improving and maintaining processes, equipment, workplaces and people.

Benefits of Implementing Green Productivity

The implementation of Green Productivity will bring about immediate as well as long-term benefits to different stakeholders. Many benefits for enterprises have been notified. These include: Reduction of waste through efficient resource utilization; Lower operational and environmental compliance costs; Reduction or elimination of long-term liabilities and clean-up costs; Increase in productivity; Compliance with government regulations; Better public image; Increase in competitive advantage; Increase in market share and profitability.

Application of Green Productivity concept to the industries brings many benefits to the employees. These include: Greater workers' participation; Potential increase in employees' share of value-added; Improvement in health and safety in the workplace; Better quality of work life.

Application of Green Productivity concept to the industries also brings benefits to the consumers. These include: High quality products and services; Reasonable pricing; on time delivery.

Conclusion

In essence, policy makers, economists and environmentalists will continue to foster this form of growth that will accelerate economic expansion in a sustainable manner. Green Productivity thus forms an integral part of the broader sustainable development agenda and moves society towards sustainable production and consumption that makes business sense. Finally, to translate the guiding principles of Green Productivity into practice, companies can adopt a systematic implementation framework. This framework encompasses the Green Productivity Methodology, and supporting Tools and Techniques.

REFERENCES

APO (2000). *Green Productivity in Asia: APO's Demonstration Projects 1995-1999.* Asian Productivity Organization.

Cascio, J. (ed.). (1996). *The ISO 14000 Handbook.* Milwaukee: ASQ Quality Press.

Culaba, A. B. and M. R. I. Purvis. (1999). A Methodology for the Life Cycle and Sustainability Analysis of Manufacturing Processes. *Journal of Cleaner Production.* 7: 443-445.

De Simone, L. D. and F. Popoff. (1997). *Eco-efficiency : The Business Link to Sustainable Development.* World Business Council for Sustainable Development.

Foust, T. D. and D. D. Gish. (1996). Future Perspective. In: Curran, M. A., (ed). *Environmental Life-Cycle Assessment.* New York: McGraw Hill.

Giarratano, J. and G. Riley (1998). *Expert Systems: Principles and Programming.* Boston: PWS Publishing Company.

Heijungs R, (ed). (1992). *Environmental Life-Cycle Assessment of Products: Guide—October 1992.* Report 9266. CML (Centre of Environmental Science) in Leiden, TNO (Netherlands Organisation for Applied Scientific Research) in Rotterdam.

Ong, Victoria D. (1997). Design and Development of a Microcomputer based Productivity Diagnostic Prototype Utilizing Embedded Rule-based Expert System. MS Thesis. University of the Philippines.

Porter, M. E. and C. van der Linde (1995). Green and Competitive. *Harvard Business Review.* 120, 122, Sept-Oct 1995.

Porter, M.E. and C. Van der Linde (1995). Green and Competitive. *Harvard Business Review.* 120, 122, Sept.-Oct. 1995.

The Netherlands ISO (1997). *ISO 14040: Environmental Management - Life Cycle Assessment - Principles and Framework.* International Organization for Standardization.

UNEP. (1993). Environmental Management in the Electronics Industry: Semiconductor Manufacture and Assembly. Technical Report No. 23.

Watanabe, S. (1993). *Resource Productivity as a New Measure for Industrial Performance.* Sony.

WCED. (1987). *Our Common Future.* World Commission on Environment and Development. Oxford University Press.

2

GREEN PRODUCTIVITY PRACTICES OF SMALL AND MEDIUM ENTERPRISES

S.N.R. Srikumar and *N. Manjula*

Introduction

The coming generation presents un-precedented challenges and opportunities. During the next 40 years or so, that is, between 1990 and 2030, the world's population is projected to grow by 3.7 billion. This would mean that the food production will need to double, and industrial output and energy use will probably triple worldwide and increase five-fold in developing countries. This growth brings with it the risk of appalling environmental damage.

Environmental problems can, and do, undermine the goals of development. There are two ways in which this can happen. First, environmental quality is itself part of the improvement in welfare that development attempts to bring. If the benefits from rising incomes are offset by the costs imposed on health and the quality of life by pollution, this cannot be called development.

Second, environmental damage can undermine future productivity. Soil that are degraded, acquifers that are depleted and eco-systems that are destroyed in the name of rising incomes today can jeopardise the prospects for earning income tomorrow.

These days, just about everyone in business is burdened with some kind of environmental challenge, from properly disposing of toxin-laden wastes to cutting landfill disposal costs to simply educating employees and customers not to toss out things that can be recycled. Nearly every company, it seems, is saddled with pressures from government, regulators, environmentalists, customers and others to change their products and processes to better accommodate these earth-sensitive times. And those pressures seem to get stronger every year.

The fact is, businesses affect the environment in atleast two ways every working day - consume energy and other resources, and create wastes that must be disposed of. Every industrial worker, from blue-collar to white-collar, does these two things. It is a fact of life for every one in every sector. And that makes everyone in business a party, however unwittingly, to environmental degradation.

But it is how one does these two things that can make a difference to the earth as well as to a company's environmental record and its productivity and profits. In coming years, the distinction between being a "green" or "ungreen" company may become a critical factor in how the stakeholder - be it stockholder, regulators, environmentalists, competitors, employees, or the consuming public - view the company's image.

Ultimately, it could determine the company's ability to remain competitive in the twenty-first century.

ROLE OF SMALL AND MEDIUM ENTERPRISES IN INDIA

As industrialisation gathered momentum so did the increase in small-scale industries. Small units play an important role in the Indian economy, as they are labour intensive and create job opportunities.

As per the Abid Hussain Committee's recommendations on Small-Scale Industries the Government of India has, in March 1997, further raised investment ceiling to Rs.3 crores for small-scale industries and to Rs.50 lakhs for tiny units. These SMEs (Small and Medium Enterprises) offer a higher productivity of capital than capital intensive enterprises, as they have low investment per worker. They help in disposal of industries, rural development, and the decentralisation of economic power. All this is required to increase and disperse economic growth.

In addition, small companies support enterpreneurial skills, stimulate personal savings, and help in developing innovative and appropriate indigenous technology, providing dynamism and contributing to competition (Rajendran, 1989). Therefore these industries are supported by the government and have been actively encouraged; no public or private enterprise with more than 100 employees has been allowed to go out of business. Several policy initiatives and procedural simplifications have been undertaken by the government to support this sector, not only for employment generation but also to enhance their competitive strength. The government has also provided measures such as greater infrastructural support, more and easier availability of credit, lower rates of duty, technology upgradation, assistance to build entrepreneurial talent, facilities for quality improvement, and export incentives (Parthasarathy, 1996).

Contributions of Small-Scale Industries (SSTs) to India's industrial production, exports and employment are significant. About 3 million SSI units employing nearly 16.7 million persons account for 35 per cent of India's total exports and 40 per cent of industrial manufacture.

Table 2.1 : Export of SMEs

Year	*Exports (At current prices)*
1994-95	29,068 (14.86)
1995-96	36,470 (25.50)
1996-97	39,249 (7.61)
1997-98	43,946 (11.97)
1998-99	48,979 (10.2)
1999-2000 (P)	53,975 (10.2)

P - Provisional

In real terms, the small-scale sector recorded a growth rate of 10.1 per cent in 1994-95 as against 7.1 per cent in 1993-94 and 5.6 per cent in 1992-93. By the year 2025, if not controlled this sector will grow even more rapidly (Parthasarathy 1996).

Table 2.2 : The growth rate of small-scale sector

Year	*Target*	*Achievement*
1991-92	3.0	3.1
1992-93	5.0	5.6
1993-94	7.0	7.1
1994-95	9.1	10.1
1995-96	9.1	11.3
1996-97	*	8.43
1997-98	*	7.7
1998-99	*	8.16
1999-200 (P)	*	8.90

P - Projected (April-December)
* Target not fixed at constant prices

The government's prime role has been to encourage growth of these industries, often neglecting environmental considerations. Industrial effluent largely comes from these 3 million small and medium sized units that are scattered throughout the country; particularly in the production of paper, sugar, leather, and chemicals. Unfortunately, only about half of the medium to large-scale industries have partial or complete effluent treatment.

Why Green Productivity in Small and Medium Enterprises

Small and medium sized enterprises are limited in their capabilities to do well in environmental performance. These firms in general have little financial, technical and manpower capability to implement adequate environmental measures. Furthermore, the public at large exerts substantially less pressure on SMEs than on large firms in the environmental, health and safety issues. As a consequence, SMEs are comparatively less active in environmental programmes. Hence, it becomes the responsibility of the Indian government to attack SMEs to practice green productivity measures to improve their environmental performance.

The Approach of Green Productivity

Unlike most of the known techniques adopted by enterprises for improvement, the approach of green productivity does not first focus on environmental issues alone; it also addresses issues pertaining to productivity. In fact when green productivity is implemented, enterprises would experience productivity improvement through reduction in expenditure on environmental protection through source reduction, waste minimization, pollution prevention and cleaner production. Hence, companies can now achieve higher productivity while protecting the environment that will enable sustainable development.

Fig. 2.1 Green Productivity

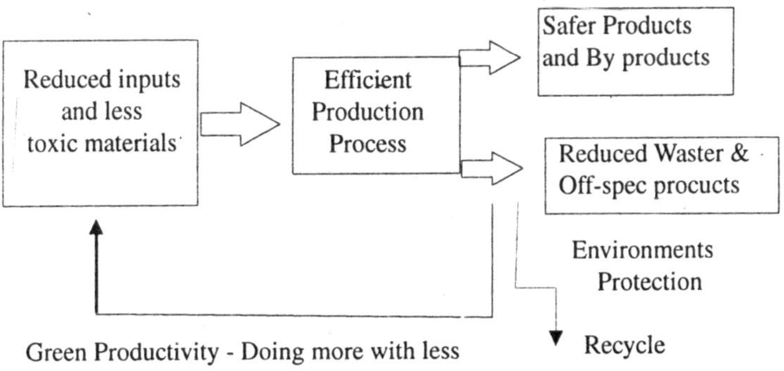

Green Productivity (GP) is defined as a strategy simultaneously enhancing productivity and environmental performance for over-all socio-economic development. This strategy is executed through the application of appropriate techniques, technologies and management systems to produce environmentally compatible goods and services.

Fig. 2.2 : Role of GP in business operation

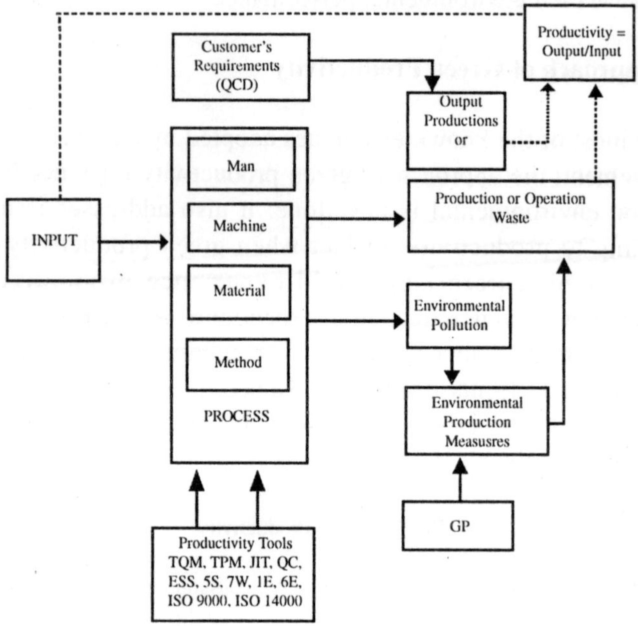

This concept was coined by the Asian Productivity Organisation (APO) in 1994 to raise public awareness on the environmental problems faced by the world today. APOs main objective is to demonstrate that environmental protection and productivity improvement can be profitably organised, even in small and medium enterprises. This is because more often than not the production processes of the small and medium enterprises are inefficient. Their production processes often lead to wastage of resources like raw materials and energy, which generate a heavy burden to the environment. Therefore, APO implemented a series of green productivity programmes for National Productivity Organisations (NPOs) to put into practice the concept of GP. Essentially, the implementation of green productivity is to achieve higher productivity and at the same time provide sustainable development. Sustainable development is defined as development that meets the needs of the present generation without compromising the ability of future generations to meet their own needs. This includes the use of services and products which meet our basic needs and improve the quality of life. The entire lifecycle of these products and services must be based on minimizing the use of natural resources and toxic substances that result in emissions.

THE GREEN PRODUCTIVITY METHODOLOGY

Getting Started

The begining of GP is marked by formation of GP team to gain baseline formation and identify problem areas. It is vital to ensure that adequate manpower are available for successful implementation.

Planning

Using the information gained in the walk-through survey along with number of analytical tools like material balance, benchmarking, ecomapping problems are identified. Following this objectives and targets are set to address the problem areas.

Generation and Evaluation of GP Options

This stage fosters to development of options to meet the objectives and targets formulated. It involves both a review of pollution prevention and control procedures and development of new options. Options are screened and prioritized in terms of their economic and technical feasibility and their potential benefits.

Implementation of GP Options

In the implementation of GP option the execution of preparatory steps include training, awareness building and competence development. Implementation is done through installation of equipment and systems along with operator instruction and hands on training.

Monitoring and Review

It is vital to check the selected GP options whether they are producing desired results. This involves monitoring the over all GP system to ensure that it is proceeding in the right direction and that targets are being achieved as per the implementation plan.

Sustaining GP

In the light of findings of GP evaluation, corrective actions can be taken to keep GP progressing towards the target.

In some cases targets and objectives themselves will have to be modified. An effective feedback system should be implemented so that new problem and challenges are dealt with. In this way GP cycle will loop back to the relevant step to implement process of continuous improvement and ensure the continuing relevance and effectiveness of process.

GP TOOLS AND TECHNIQUES

GP tools and techniques includes clues and directions for generation of options and enable their systematic implementation. Different tools and techniques are used at different stages of GP implementation process.

At the start of *GP Process Flow Charts* and *Process Flow Diagrams* provide a graphical means representing activities, processes and material flows. A *material balance* bus on a process diagram allows quantitative assessment of material inputs and outputs.

Benchmarking : Is often used to identify gaps in performance of comparing the current achievement of a department or a company against what others have done.

Brainstorming and Concentration : Help in identification of possible root causes of problems and for data collection.

ISHIKAWA and Effect Analysis : Is used to uncover the reasons behind problems and gives graphical representation of causes and effects.

Ecomapping : A simple and practical visual to provide a bird's eye view of company's operation and thus a quick inventory of practical problems.

Pareto Analysis : Graphical tool used to isolate key problems that causes most significant impact.

Check Sheets : are used for collecting data over time to show trend of recurring problems.

Control Charts : are used to show deviation and variability in performance. GP options are assessed using *Brain Storming* and *Cost Analysis* which facilitates the comparison of alternatives in terms of the monetary cost involved and the benefits that can be obtained.

Feasibility Studies are often at par with audits for selection of alternative options. As options are implemented *spider-web* diagram will provide a visual way of showing progress and performance against several target at once.

GP techniques are used to bring about changes that will result in better environmental performance and approved productivity. They range from simple house keeping techniques to designing products.

Good House Keeping : GP techniques includes awareness programmes and with 55 management techniques keeping processes, equipment, workplace and workforce organise it. Further prevent waste conserve and save energy and enhance operation of organizational procedure.

Design Change : The environmental impact of a product is to a large extent determined by design. The company can minimize the negative impact of products and process on the environment if in accounts for designing environmentally compatible product as in very beginning from product planning, design and development.

Process Modification : This is a key GP technique which encompasses both simple and complex change from replacing inefficient or old processes with new technology or totally changing production process.

Waste Management

Waste stream segregation and the promotion of recycling, reuse and recovery are two broad techniques used to reduce the account of waste a company produces to improve waste disposal. Substantial improvements can be made in the nature and quality of waste produced by purification of material inputs.

Benefits of Implementing Green Productivity in SME

The implementation of green productivity will bring about immediate as well as long-term benefits to different stakeholders.

For Small and Medium Enterprises

- Reduction of waste through efficient resource utilization.
- Lower operational and environmental compliance costs.
- Reduction or elimination of long-term liabilities and clean-up costs.
- Increase in productivity.
- Compliance with government regulations.
- Better public image.
- Increase in competitive advantage.
- Increase in market share and profitability.

For Employees

- Greater worker's participation.
- Potential increase in employees' share of value added.
- Improvement in health and safety in the workplace.
- Better quality of work life.

For Consumers

- High quality products and services.
- Reasonable pricing.
- On time delivery.

Difficulties in Promoting GP among SMEs

Although GP has many benefits, a number of factors may actually inhibit putting such programme into effect. These factors fall mainly into two categories : "Technical" and "Financial".

Technical barriers work to impede the ability of a company to develop, evaluate and implement GP programmes. These barriers include limited awareness of pollution management issues at company's decision-making level, lack of inhouse expertise on GP and the absence of readily available GP technologies that can be adopted directly. In addition, attitude towards changing established industrial processes or practices may be negative, and such attitude tends to block new ways of preventing pollution.

Pollutant generators may be reluctant to take risks with new, unproved technologies or to compromise other business goals and practices; they may also distrust alternative processes or simply be uninterested in changing their habitual ways of doing business.

The unavailability of capital for plant modernization often becomes a significant obstacle to implementing GP even though GP measures may lead to cost savings. Major companies may have capital to upgrade inefficient processes, but small and medium firms often do not.

Other Obstacles for SME's in India to Implement GP Include

- Lack of cost accounting system to identify costs of environmental control and benefits of source reduction.
- Substantial portion of SMEs currently running without proper registration and licensing, resulting in lack of authority from Government to deal with environmental issues associated with these firms.
- Lack of aggressive enforcement of pollution control regulations in the country; industries are thus less likely to invest in GP if they are allowed to illegally dispose of waste.
- Usually consisting of a small staff, SMEs are generally lack of manpower to acquire for new technical and management skill required for GP implementation.

GP DEMONSTRATIONS

A Case Study of Tanchem Industries Meerut and Nasser Tannery, Vaniyambadi

About the Companies

Tanchem Industries, Meerut is a small-scale tannery processing about 200 buff calf hides (or an equivalent quantity of cow or buffalo hides) per batch. In a year, it processes about 48 batches and employs around 30 workers in all. The unit, which is located in the northern part of India, produces nappa, oil pull nappa and other varieties of leather used by the garment and shoe manufacturing industries.

Nassau Tannery Company, Vaniyambadi is a medium-scale tannery that processes about 2500 sheep/goat skins per batch. In a year, it processes around 300 batches. The unit, which is located in the southern part of India, is 100 per cent export-oriented and produces suedes, nappa, nubuck and other varieties of leather for garment and golf glove manufacture. It employs around 175 employees and has an annual turnover of about Rs. 2 billion (US$ 45 million).

These two companies represent between them many of the main characteristics of the tannery industry in India and allow an assessment of how issues of scale affect environmental and economic performance in the sector. Moreover they allow a comparison of the different leather production processes used for hides and skins.

Why GP in Tanneries?

Indian tanneries are generally outdated, use obsolete technology and depend excessively on traditional skills. This means that their productivity is well below the norm for tanneries in developed economies such as Europe. However, exports from the leather sector constitute seven percent of India's export basket, making it one of the nation's top five export earners. Now, in the wake of the globalisation of the Indian economy and the liberalization of economic and trade policies, the industry is poised for further growth to achieve a greater share in the global market. Environmental excellence coupled with much-needed productivity improvements will be vital to fully capitalize on this opportunity.

Liquid wastes comprise the vast majority of pollution from tanneries. In general, these wastes have high biological oxygen demand (BOD), high concentrations of suspended solids, dissolved solids, oils and grease. Effluent may also contain chrome, sulphide, nitrogen and other chemicals. Improperly treated tannery waste water has severe impacts on surface water, ground water and soil quality and there have recently been mass closures of tanneries in the State of Tamil Nadu in response to this problem.

GP is, therefore, vital to promote the establishment of environmentally friendly factories and to show that environmental protection and profitability can be harmonized particularly in SMEs. To implement GP the companies used the methodology described earlier.

Main Issues

The production of finished leather from raw hides and skins can be grouped under three major operations. Beam house operations; tanning operations; and, post-tanning operations. All aspects of these operations were found to be potential sources of waste generation, although the beam house operations - where the raw hides are cleaned and prepared - was found to contribute over 50 per cent of the tanneries' total pollution load.

As expected water pollution was found to be the main problem, with large volumes of waste with high BOD, COD, TSS and TDS produced from each unit. The impact of these pollutants was varied, and included human health problems due to waste from the spent chrome bath, soil contamination from liming waste water and ground water pollution from soaking wastewater.

Causes of this pollution included: process inefficiencies (e.g. the problem of unfixed fat liquors due to the use of cold water); spillages (e.g. chemicals at the neutralization stage); unoptimised and inefficient processes (e.g. at the soaking and pasting stages); and a lack of accurate measuring due to a lack of necessary machinery (e.g. in the deliming, bating and degreasing stage). Poor quality control, impurities in the chemicals used, an unskilled and untrained workforce and a lack of water recycling infrastructure were also highlighted.

It was also found that a large amount of solid waste is generated from the tanneries. The major sources were unhairing, liming mud, fleshings, splitting, shavings, trimmings and buffings amongst others. However, all solid waste generated from the tannery has a market value and is sold to local vendors who come to the unit to collect it. For example, fleshings are used in the fertilizer industry, while even finished leather trims are used by cobblers for patch work. Material balances were drawn up for both units and a cost analysis was done on the waste streams they produced. The waste stream from the Buff Calf process in Tanchem Industries was calculated to cost about Rs. 2,830/batch (US$ 63) due to chemical loss and treatment costs. For Nasser Tannery the total cost of its waste stream was Rs. 8,166 per batch (US$ 182). The company consumed about Rs. 1,750,000/month (US$ 39,000/month) of chemicals.

GP Solutions and Implementation

Based on a cause analysis, material and component balance, literature survey and brainstorming among team members and experts, green productivity options were developed. A total of 45 options were developed for Tanchem Industries' hide-based process and 59 for Nasser Tannery's skin-based process. These were divided into good housekeeping, process modification, material change, elimination and reduction, equipment modification, technology change and recycle, reuse and recovery.

GP options ranged from changing the felt on the roller of the sammying and setting machine (to ensure uniform pressure and a better grip on the hides) to the implementation of 'green' fleshing immediately after soaking (to reduce the use of lime and sulphide and to allow reuse of fleshings at the end of the process). The options were assessed in terms of their implementation feasibility. Over 70 per cent of the feasible options were implemented.

Benefits of GP in Tanneries

Many of the implemented options were found to give substantial economic and environmental benefits. For example, in the hide-based process, a reduction in the use of fat liquor from 18 per cent to 8 per cent of the weight wet blue (along with some process modifications) involved little expenditure. These changes, however, reduced organic load and brought economic benefits worth Rs.175,900 (US$ 3,926.5). In the skin-based process recovery of chrome from the spent tanning bath and the tanning wash resulted in less chrome load to the environment, the elimination of chrome sludge from the ETP and a cost saving of Rs. 727,000 (US$ 16,235) per year. Given an initial investment of Rs. 316,000 (US$ 7,060) this translates into a payback period of less than six months.

Overall water consumption dropped by 2,000 litres per batch of 850kg buff calf hide and by 20,000 litres per batch for skins. Both the hide-based and skin-based processes' wastewater showed reductions of BOD, COD and TDS after the implementation of the GP (for details see Table 2.3).

Table 2.3 : Improvements in the Quality of Wastewater

	Tanchem	*Nasser*
Reduction in COD Load (% Reduction)	19.1 kg/batch (10%)	157.2 kg/batch (17.2%)
Reduction in BOD Load (% Reduction)	3.7 kg/batch (5.8%)	31.8 kg/batch (15.1%)
Reduction in TDS Load (% Reduction)	5 kg/batch (1%)	1046.8 kg/batch (40.5%)

One of the major achievements of implementation of GP solutions was that the marketability of products from Tanchem Industries was found to have increased by over 4.3 times. The unit's productivity also rose to about 210 batches per annum.

Implication

The GP programmes at Tanchem Industries and Nasser Tanning Co. showed that the application of creative thinking can result in financial benefits for a factory while allowing it to comply with regulatory requirements. The implementation of GP at the two sites also showed that a step-by-step programme could provide an effective way for both small, and medium-sized enterprises to generate green productivity options.

A Case of Dintex Dyechem Ltd., Ahmedadad

About the Company

Dintex Dyechem Ltd. is located in the Vatva Industrial Estate in Ahmedabad in Gujarat State, India. Its main products are dye intermediates such as vinyl sulfone, 6-Nitro, 1,2,4 acid and SPMP as well as reactive, acid and direct dyes. It is a medium sized company that has about 225 employees and a turnover of some 200 million rupees (US$ 4,464,200).

Why GP in Chemical Industries

Business Sustainability

The Indian dyestuff industry has a poor environmental performance record. Hundreds of factories in the sector have been closed in recent years for pollution infringements. Dyestuff companies now face the challenge of improving their performance to respond to both tightening regulatory enforcement and the demands of an international market that requires good environmental practice. By implementing GP, Dintex hoped that it could both improve its own performance and demonstrate to the rest of the Indian dyestuff industry that improvements are technically and economically feasible.

Product Improvement

The GP demonstration project at Dintex Dychem focused on the production of vinyl sulphone. This chemical is one of the major dye intermediates exported to developed countries. It generates a revenue of US$ 64 million a year for India. However, manufacturers are facing acute environmental problems associated with vinyl sulphone production. Dintex hoped that a GP approach would help it to resolve this challenge. To implement GP, the company used the methodology described earlier.

Main Issues

Vinyl sulfone is produced in a batch process that takes about five days to complete. The four major stages in the production process are sulfonation, reduction, condensation and esterification. Equipment includes boiler and chilling plant and a thermic fluid heater.

A breakdown of the key stages and main inputs and outputs are illustrated in the process flow diagram given below:

Fig. 2.3 : Process Flow Diagram for Vinyl Sulfone Production

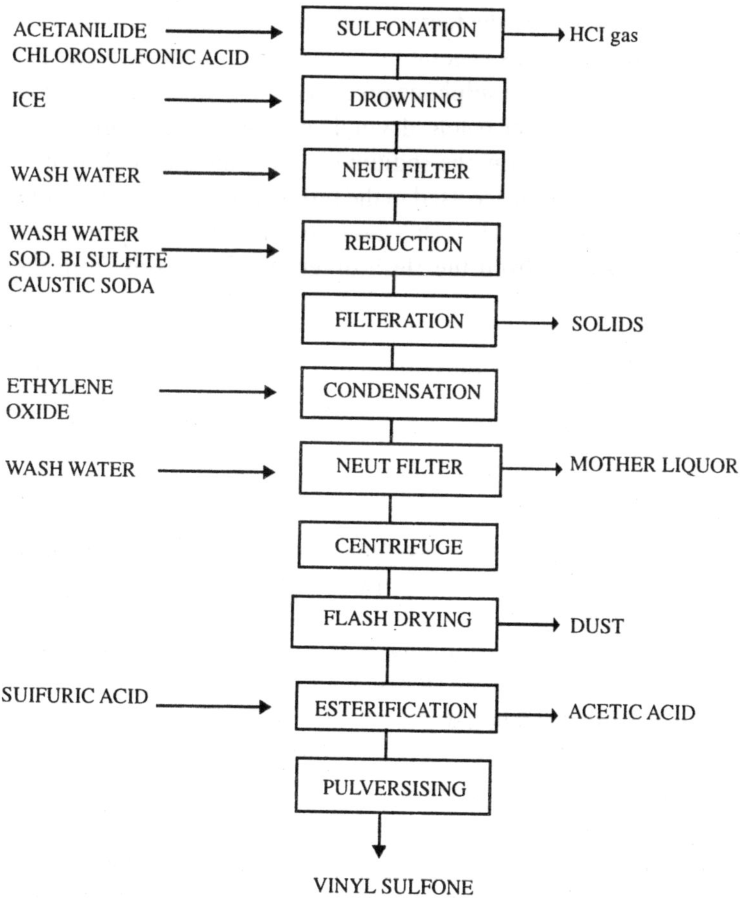

The review of Dintex's manufacturing process highlighted the following main issues:

- The plant's electrical systems were generally under-loaded and not operating at optimal efficiency. Its steam distribution system was poorly maintained and leaky and its refrigeration system was inefficient.
- The plant's waste water treatment system - which comprises a neutralization tank, Neutch filter and clariflocculator and aeration, sedimentation and polishing tanks was not adequate for the entire pollution load generated by the factory and needed to be modified.
- The plant was generating over 18.6 m^3 of waste water per tonne of vinyl sulphone produced; 80kg of solid waste was generated per batch of vinyl sulfone (this comprised the impurities from the sodium bisulfite and insoluble matter present in the ice used in the production process). Gaseous emissions included HCl gas from the sulfonator, particulate emissions from the flash dryer and flue gases from the boiler.

The plant's pollution outputs, inefficiencies and losses were attributed to:

- Poor raw material quality (e.g. impure sodium bisulphide and ice) and improper storage.
- Poor training, housekeeping, operational and maintenance negligence, manual quantity estimation and poor monitoring.
- Inadequate technology (e.g. lack of hood on dumping vessel; inadequate HCl recovery system) and poor process and equipment design.

GP Solutions/Implementation

As a result of the process review 66 potential improvements were highlighted. 36 were found to be technically and economically feasible, although half of these needed trials or laboratory studies before they could be implemented.

A number of different pollution treatment schemes were also analysed, however it was found that treatment was not economically feasible and would generate large quantities of hazardous solid waste which would itself pose handling and disposal problems. Instead it was found that in-process pollution reduction and resource recovery represented the best options.

The improvements implemented included:

- Enhanced housekeeping, maintenance and monitoring. For example, the factory's steam distribution system was insulated and better process controls, such as temperature monitoring devices, installed.
- Pollution reduction through the retrieval and recycling of resources from waste. For example, sulfanilic acid was recovered from the mother liquor and salt from waste liquor. Acidic mother liquor was retrieved for subsequent use in alum production in the paper industry.
- New technology development: For example, the HCl scrubber design was modified by installing a packed-bed type secondary stage to improve HCl recovery. The cake washing mechanism was also improved to reduce the amount of wash water needed.

Benefits of GP

The implementation of the improvements outlined above generated substantial environmental and economic benefits. These included:

Material Savings

Resource recovery per batch:

- 224 kg HCl.
- 200 kg sulfanilic acid.

- 8,800 kg acid.
- 7,200 kg of salts.
- 82 kg condensation product (equivalent to about 100 kg of vinyl sulfone).

Economic Savings

This resulted in savings of over Rs.16 million (US$ 357,136) a year - from an initial investment of some Rs. 17 million (US$ 379,457).

Implication

The Dintex GP project showed that a pollution prevention approach to environmental improvement can give significant benefits both in terms of production efficiency and profitability. The GP survey made it clear that the waste generated by dyestuff manufacturing requires complex and expensive treatment that produces many problems of its own. It also made it apparent that many of the outputs of the manufacturing process can be reused and thus represent a resource and income stream. In the highly competitive world dyestuffs market, Dintex's GP team concluded that any company that reduces it pollution load through a systematic programme of pollution prevention and resource recovery will gain a substantial competitive edge.

Conclusion

The major environmental concerns in India today are poverty coupled with growing population and the side effects of enhanced industrial activities. As long as poverty remains the main stumbling block, industrialization provides hope of significantly improving the standard of living. One of the measures most talked about that might gain recognition within these industries is, sustainable development. Removal of poverty and environmental protection are two sides of the same coin that is sustainable development

(Dwivedi & Khator, 1995), but policy makers, governments, politicians, and industrialists have challenged many of the underlying values and assumptions of sustainability. Sustainability or sustainable development can also be described as development or progress that meets the needs of the present without compromising the ability of future generations to meet their own needs.

Although, industrializaton is seen as a solution to providing economic growth and increasing employment levels, irrespectively, industries, whether large or small, low-tech or hitech, manufacturing or agricultural, all inevitably produce discharges and wastes that are capable of polluting. Where high population and economic growth demands resources (inputs) and discharges (output) in the form of pollutants, not many industries have arrived at suitable suggestions on sustainable measures, thus putting pressure on environment.

However, the suggestion that learning to balance ecological principles, economic growth and social responsibility be priorities of business (JOHANNSON, 1994) does eventually make more sense.

Sustainable development through GP challenges industry to produce high levels of inputs and generating less wastages with a more effective use of raw materials in production that would eventually result in diminishing costs. This greener corporate image could then lend to an increased market share.

REFERENCES

Anshuman Khare, "Greening the Earth is Corporate Responsibility", Indian Management, May 1998, pp.46.

Dr. C.M. D'Souza, Latrobe University, Australia, "Integrating Environmental Management in Small Industries of India", *Electronic Journal*, issue: 14.

Dr. Gitika Kapoor, "Strategies for Sustainability - The Organisation's Agenda, Indian Management, June, 1999.

Gupta, M. & Sharma, K. (1996, third quarter). Environmental operations management. An opportunity for improvement. *Production and Inventory Management Journal*, 40-46.

Hart, S. (1997, Jan./Feb.). Beyond greening : Strategies for a sustainable world. *Harvard Business Review*, 67-76.

Jitendra Kumar Das, "Responding to green concerns: The roles of government and business, *Vikalpa*, Vo1.27, No. 1, January 2002, p.3-11.

Johannson, L. (1994). Profits from green ventures. *The Management Accounting Magazine* (CMA), 68(7), 17-21.

Major, R.N. Wahal, "Environmental Business Management", Indian Management, August, 1998, pp.57-61.

Mani, M., Pargal. S., & Huq, M. (1996). Does environmental regulation matter? Determinants of the location of new manufacturing plants in India in 1994 (World Bank Policy Research Working Paper # 1718). Retrieved from the World Wide Web *ho://www.worldbank.or.&ipr/work_paper/index.htm.*

Mathew, P.M. 2004, "Role calls for relook", The Hindu Survey of Indian Industry, 2004, p.331.

Mukul P. Gupta, "Is Marketing responsible for Environmental Crisis?," Indian Management, 1998, pp.53-57.

Nyati, K. (1988). *Problems of pollution and its control in small scale industries.* New Delhi, India: Friedrich Elbert Foundation.

Pargal S., Mani, M., & Huq, M. (1997). Inspections and emissions in India : Puzzling survey evidence on industrial water pollution (World Bank Policy Research Working Paper #1810). Retrieved from the World Wide Web *hM://www.worldbank-ory./niRL/work_paper/index.htm.*

Parthasarathy, S. (1996). *Whither small scale industry? In The Hindu Survey of Indian Industry,* (pp.14-26). Madras, India: Kasturi & Sons, National Press.

Peattie, K. (1995). *Environmental marketing management : Meeting the green challenge.* London: Pitman.

Porter, M. & Linde, C. (1995, Sept./Oct.). Green and competitive: Ending the Stalemate. *Harvard Business Review*, 120-134.

Rajendran, S. (1989). *Institutional support to SSIs.* The Hindu Survey of Indian Industry, M/s (pp.279283). Madras, India: Kasturi & Sons.

Shaman, D. (1996). *India's pollution regulatory structure and background.* U.S.A.: The World Bank. Retrieved from the World Wide Web *: hqg://www.worldbank.or-aJnipr/work /nipr/work naner/index.htm.*

Shen-Yann Chiu, "Applications of corporate synergy systems for promoting green productivity in small and medium enterprises in Taiwan," APO Formum, November, 2001.

3

ENVIRONMENT FRIENDLY FORCING SYSTEM IN TEXTILE INDUSTRY

V.B. Upadhye and *P.R. Wadje*

Introduction

The growth of Indian cities has been attaining alarming rate in the past few decades. The increasing human needs and the technological advancements coupled with population explosion has created a situation such that if we do not have measures to prevent the environmental damage by interaction, our survival itself would become difficult. Environmental pollution is assuming dangerous proportions throughout the globe and India is not free form the disastrous effects of pollution. Industrial growth significantly contributes towards economic growth. However, industrial progress brings along with it a host of environmental problems. Many of these problems could be avoided if industries are located on the basis of environmental considerations. Injudicious siting of industry can seriously affect the environmental features such as air, water, land, flora, fauna, human settlements and health of people. The entrepreneur should be fully aware of these implications and he should take necessary steps while setting up the industry so as to minimize the possible adverse effects on the environmental resources and quality of life. Poverty and the extreme backwardness of the people are our biggest pollutants. The developing countries are too much infatuated with the Western model of development. The result has been " the whole development process has not been built on the foundations of a traditional society but fallen on it like a nuclear bomb and blown it apart-eco-development is no more than a sensible pre-condition of sustainable development." But poor countries like India need two-stage two-pronged attack. They are to eliminate the contamination and poverty. This should mean safe drinking water and disposal of human waste to control diseases, food to eradicate malnutrition and medicare to deserving people. India is suffering from air, water and noise pollution contributing to the overall degradation of the environment. Today conservation of resources plays a key role to make development compatible with environment under unprecedented demographic pressure and rapid pace of development. Textile industry is one of the largest industries in India and pointed attention is being directed towards pollution of water, aquatic media and air caused by the textile mills.

Sources of Pollution and Effects of Pollination

Water pollution caused by the textile mills is mainly attributed to the various waste streams coming out to the processing operations like sizing, disizing, scouring, bleaching, mercerization, dyeing, printing and finishing. Moreover, water discharged from the water treatment plants also contribute to the water pollution. About 50 per cent of the textile mills are using 200-300 litres of water per kg. of the fabric produced while about 20 per cent of the mills are using below 200 litres per kg. of the fabric produced and the rest are using 300 and more litres of water per kg. of the fabric processed. Of the total water consumed in a typical textile mill, 40 per cent will be consumed in bleaching and finishing, 15 per cent in boilers, 15 per cent in humidification in spinning and weaving departments and 7 per cent in domestic and sanitary purposes.

Caustic alkalinity from kiers, free chlorine from bleaching wastewater etc., is toxic and harmful. These make water unfit for use, destroy vegetation and crops. The *p*H of receiving streams will be very high. If *p*H exceeds nine or falls below five, it will have adverse effect on aquatic biota especially invertebrate life in water.

Chemically reducing agents, biologically degradable chemicals and surface-active agents deplete oxygen in receiving streams.

Sulphur dyes, sodium hydrosulphite, sodium nitrite, sodium sulphide and Rogalite will deplete the dissolved oxygen of the receiving waterbody because of their reducing characteristics. Sizes from gray fabric and other organic solids present in the effluent will consume dissolved oxygen from water for their biochemical stabilisation. Oily scum formed by the surface active agents like greases, kerosene and detergents on the surface of the water will interfere with the mechanism of oxygen transfer at the air-water interface. The effluent discharged with high temperature will increase the temperature of the receiving body, thereby reducing the solubility of oxygen in water. There are some inert pollutants in suspension or in dissolved state, which are harmless but can be detrimental if present in large amounts. They can blanket the streambed by way of deposition thereby destroying the plant and invertebrate life. The soluble dyes and colours present in the wastewater, because of which the photosynthesis reaction carried out by some micro organisms in the biological treatment of the sewage are hampered. The dissolved solids of the effluent can clog the porous structure of the soil on which the effluent is discharged. Intake of air from the atmosphere by plants and vegetations on such soil will be hampered and subsequently their growth also will be retarded. Some of the salts deposited on the soil, from the effluent, by evaporation of water due to natural drying can alter the osmotic pressure balance across the plan roots. This may further lead to adverse effects on the absorption of water and other nutritious compounds from the soil by the plants.

Many of the chemicals used in textile processing are toxic. Sodium hypochloride is used as a bleaching agent on a large scale. Chlorine acts as a sterilizing and decolorizing agent by virtue of its ability to oxidise organic coloring matter. It has been shown that brief exposure to air containing 3000 mg/cu.m.of chlorine kills even large animals. The sensitivity of man to chlorine varies greatly. Exposure to 10 to 20 mg/cu.m.of Cl_2 causes a stringing or burning sensation in eyes, nose and throat. Sometimes causing headache. There may be redness and watering of eyes, sneezing, coughing or loss of voice. Bleeding of nose may occur. Exposure for 30 to 60 minutes to 40 to 60 mg/cu.m. cannot be borne by man for more than one minute. Sulphuric acid is used in wet processing. Concentrated sulphuric acid attacks the skin instantaneously producing burns, which take a long time to heal. Even after healing of wound, permanent scars are left behind. Sulphuric acid mists are strongly irritant and the inhalation of about 3 mg/cu.m.causes choking sensation. Persons accustomed to the exposure are unable to notice concentration of this order of magnitude. Following is the summary of the effects caused by various chemicals.

Chemical	Use	Effects
1. Caustic Soda	Bleaching, mercerising dyeing, scouring etc.	Direct contact - irritation of nasal tissues, sneering.
2. Pentachlorophenol	Sizing and Printing.	prolonged or repeated contact with skin cause dermatitis and systemic intoxication. Injuries to lungs, liver, kidney, pancreas and spleen may occur.
3. Carriers dichloro or trichloro Benzene derivates	Dyeing and Printing	Toxicological effects of dichloro benzene is primarily injury to liver and secondarily to kidneys. Short exposure at high concentration may result in depression of central nervous system.
4. Dyeing with azoic colours	Dyeing/printing.	Evolution of N_2 gas. It is irritant to eyes, nose and upper respiratory mucosa. In concentrations of 100 to 1000 ppm or more it causes deaths of cats, mice, pigs, rats and rabbits. Acute exposure to Nitrogen dioxide has resulted in death of man due to asphyxia, resulting from pulmonary odema.

Contd...

Contd...

Chemical	*Use*	*Effects*
5. Hydrocholric acid	Dyeing	75 to 150 mg/cu.m.HCI gas in air causes work impossible. Prolonged exposure to 15 to 75 mg/ cu.m. causes erosion of theeth. Mists containing HCl can cause bleeding of nose and gums, ulceralin of the nasal and oral mucosa and render the skin of the face so tender that shaving becomes painful.
6. Acetic Acid	Dyeing	Skin, eye and mucous membrane or exposed teeth are affected. Oral ingestion of glacial acetic acid is very serious. Blacking and hyperke-natoses of skin of hands, conguctivitis, bronchitis, pharygitis and erosion of teeth, both incisors and canines are major effects.
7. Resin finished Formaldehyde	Finishing	At 4-5 ppm the discomfort in terms of tingling sensation in the eyes and nose and posterior pharynx increases rapidly.
8. Cotton dust	Textile Mills	Byssinosis is an occupational disease induced by inhalation of cotton dust produced during the processing of cotton, flax or hemp. Reverisble tightness of chest, dyspnea and cough with concomitant decrease in expiratory flow rates over the course of the work shift.

Noise Pollution

Noise has been described as a sound without agreeable musical quality, or as an unwanted or undesired sound. Noise is no less a pollutant than the toxic chemicals in our environment. Textile sector is a bigger source of noise. Noise mainly affects hearing, but it may also affect health in other ways too. Prolonged exposure to noise may cause a progressive symmetric bilateral impairment of hearing. The patho-physiological process which takes place in the inner ear, in the cochlear region. The disease originates with a degenerative impairment of the neuro-sensorial cells of the corti organ. After passing through several phases, this process culminates it the total atrophy of these cells. Its subsequent evolution follows more or less the same pattern as that of sensory deafness attributable to any other cause.

Air Pollution

In air pollution the global problems include C.F.C. and ozone depletion, acid rain and climate problems (greenhouse effect) sources of air pollution are: (1) Industry (2) transportation (3) agriculture (4) municipal activates (like refuse disposal, sewage treatment etc.). The problem is so serious in India and other developing countries that thousands of tones of dust, sand and obnoxious gases are making the life impossible. A healthy man intakes about 16.5kg air/day. If this air is polluted, intake of oxygen is reduced. Besides these there are adverse effects of pollution on men, animals, vegetation and materials. Corrosion of materials by acid mist and acidic gases like SO_2, CO_2, CO, NO_x, etc. can vary significantly. They are miscible in air and their separation is quite difficult. The emission from automobiles is another source of air pollution. This pollution is described as "disease of wealth". Exhaust from the vehicle in a good condition contains 3.5 per cent CO, going up to 7 per cent, when the engine is idling. In the vicinity of traffic, in India, the concentration may be as high as 25ppm. The World Health Organization has prepared a report of Mumbai city wherein it was found that the single most contributing factor (55 per cent) towards air pollution is auto emission. Indusrty contributes 30 per cent of the air pollution, while construction activities and garbage on streets contribute to about 15 per cent. Teraethyl lead is added to petrol to increase its octane number. The present concentration of lead in petrol is 0.560gm/lit. The American National Sciences Academy has set a standard of 15 to 20 micrograms/100 ml of lead in the blood at a tolerable limit. The results are alarming in a city like Mumbai, wherein locations having high traffic density, the air contains 300 to 600 nanograms of lead /m^3.

Environment (Protection) Bill

A country seeking increasing presence of high technology area of industrial activity, it will recognised, cannot turn a Nelson's eye towards environmental protection. The pace of industrial development envisaged currently, it is further felt, warrants that environmental control ought to be accorded high priority lest the situation should aggravate further and become difficult to be tackle, technically as well as financially. Twenty highly polluting industries have been noticed by the Central Government as falling within the purview of industrial licensing. Specific clearance is necessary from the Department of Environment as a pre-condition for setting up an industrial unit in these lines of manufacture. In the wake of the Bhopal tragedy, caused by the leakage of poisonous gas (MIC) in pesticides manufacturing plant there, a policy paper for legislative and institutional framework to regulate the import, manufacture, handling and disposal of hazardous substances has been prepared. The Central Board is apex body attending to the problem of water and air pollution at the national level and advises the Central Government in formulating all the policies and plans for action of environmental pollution. Decisions were taken at the United Nations Conference on the Human Environment held at Stockholm in June 1972, in which India participated to take appropriate steps for the protection and improvement of human environment.

The Environment (Protection) Bill passed by the Parliament in Thirty Seventh Year of the Republic of India is called the Environment (Protection) Act, 1986 extends to whole of India.

The Environment (Protection) Bill passed by the Parliament in the Thirty Seventh Year of the Republic of India is as follows :

Chapter - I

Preliminary:

1. (1) This Act may be called, the Environment (Protection) Act, 86.
   (2) It extends to the whole of India.
   (3) It shall come into force on such date as the Central Government may be notification in the Official Gazette, appoint and different dates may be appointed for different provisions of this Act and for different areas.
2. In this Act, unless the context otherwise requires -
   (a) 'Environmental' includes water, air and land and the inter-relationship which exists among and between water, air and land and human beings, other living creatures, plants, micro-organisms and property.
   (b) 'Environmental Pollutant' means any solid, liquid or gaseous substance present in such concentration as may be or tend to be injurious to environment.
   (c) 'Environmental pollution' means the presence in the environment of any environmental pollutant.
   (d) 'Handling' in relation to any substance, means the manufacture, processing treatment, package, storage, transformation, use, collection, destruction, conversion, offering for sale, transfer or the like of such substance.
   (e) 'Hazardous Substance' means any substance or preparation which by reasons of its chemical or physico-chemical properties or handling, is liable to cause harm to human beings, other living creatures, plants, micro-organism, property or the environment.

(f) 'Occupier' in relation to any factory or premises, means a person who has control over the affairs to any substance, the person in possession of the substance.

(g) 'Prescribed' means prescribed by rules made under this Act.

Chapter - II

3. (1) Subject to the provisions of this Act, the Central Government shall have the power to take all such measures as it deems necessary or expedient for the purpose of protecting and improving the quality of the environment and preventing, controlling and abating environmental pollution.

(2) In particular, and without prejudice to the generality of the provisions of the sub-section.

(3) Such measures may include measures with respect to all or any of the following matters, namely:

(a) Co-ordination of actions by the State Government, officers and other authorities.

(i) Under the Act, or the rules make thereunder.

(ii) Under any other law for the time being in force which is relatable to the objects of this Act.

(b) Planning and execution of a nation-wide programme for the prevention, control and abatement of environmental pollution.

(c) Laying down standards for the quality of the environment in its various aspects.

(d) Laying down standards for emission or discharge of environmental pollutant from various sources whatsoever, provided that standards for emission or discharge may be laid down under this clause from different sources having regard to the quality or composition of the emission or discharge of environmental pollutants from such sources.

(e) Restriction or class of industries, operations or processes shall not be carried out to certain safeguards.

(f) Laying down procedures and safeguards for the prevention of accidents which may cause environmental pollution and remedial measures for such accidents.

(g) Laying down procedures and safeguards for the handling of hazardous substances.

(h) Examination of such manufacturing processes, material and substance as are likely to cause environmental pollution.

(i) Carrying out and sponsoring investigations and research relating to problems of environmental pollution.

(j) Inspection of any premises, plant, equipment, machinery, manufacturing or other processes, materials or substance and giving, by order of such directions to such authorities, officers or persons as it may consider necessary to take steps for the prevention, control and abatement of environmental pollution.

(k) Establishment or recognisation of environmental laboratories and institutes to carry out the functions entrusted to such environmental laboratories and institutes under this Act.

(1) Collection and dissemination of information in respect of matters relating to environmental pollution.

(m) Preparation of manuals, codes or guides relating to the prevention, control and abatement of environmental pollution.

(n) Such other matter as the Central Government deems necessary or expedient for the purpose of securing the effective implementation of the provisions of this Act.

4. The Central Government may, if it considers necessary or expedient so to do for the purpose of this Act, by order, published in the Official Gazette constitute an authority or authorities by such name or names as may be specified in the order for the purpose exercising and performing such of the powers and functions (including the power to issue

direction under section (5) of the Central Government under this Act and for taking measures with respect to such of the matters referred in sub-section (2) as may be mentioned in the order and subject to the supervision and control of the Central Government and the provisions of such order, such authority or authorities may exercise the powers or perform the functions to take the measures so mentioned in the order as if such authority or authorities had been empowered by this Act to exercise those powers or perform those functions or take such measures.

5. (1) Without prejudice to the provisions of sub-section (3) of section 3, the Central Government may appoint officer with such designations as it thinks fit for the purposes of this Act and may entrust to them such of the powers and functions under this Act as it may deem fit.

(2) The officers appointed under sub-section (1) shall be subject to the general control and direction of the Central Government or if so directed by that government, also of the authority or authorities, if any, constituted under sub-section (3) of section 3 or of any other authority or officer.

6. Notwithstanding anything contained in any of the law but subject to the provisions of this Act, the Central Government in the exercise of its powers and performance of its functions under this Act, may issue directions in writing to any person, officer or any authority and such person, officer, or authority shall be to comply with such directions.

Explanation

For the avoidance of doubts, it is hereby declared that the power to issue directions under this section includes the power to direct —

(a) The closure, prohibition or regulation of any industry, operation or process.

(b) Stoppage or regulation or of the supply of electricity or water or any other service.

7. (1) The Central Government may, by notification in the Official

Gazette make rules in respect of all or any of the matters referred to in section 3.

(2). In particular and without prejudice to the generality of the foregoing power, such rules may provide for all or any of the following matters, namely:

(a) The standards of quality of air, water or soil for various areas and purposes.

(b) The maximum allowable limits of concentration of various environmental pollutions (including noise) for different areas.

(c) The procedures and safeguards for the handling of hazardous substance in different areas.

(d) The prohibition and restrictions on the handling of hazardous substance in different areas.

(e) The prohibition and restrictions on the location of industries and carrying on of processes and operations in different areas.

(f) The procedures and safeguards the prevention or accidents which may cause environmental pollution and for providing for remedial measures for such accidents.

Chapter - III

Prevention, Control and Abatement of Environmental Pollution

8. No person carrying no any industry, operation or process shall discharge or emit or permit to be discharged or emitted any environmental pollutant in excess of such standards as may be prescribed.

9. No person shall handle or cause to be handled hazardous substance except in accordance with such procedure and after complying with such safeguards as may prescribed.

10.(1) Where the discharge of any environmental pollutant in excess of the prescribed standards occurs or it apprehended to occur due to any accident or other unforeseen act or event, the person responsible for such discharge occurs or is apprehended to occur shall be bound to prevent or mitigate the environment pollution caused as a result of such discharge and shall also forthwith —

(a) Intimate the fact of such occurrence or apprehension of such authorities or agencies as may be prescribed.

(2) On receipt of information with respect to the fact or apprehension of any occurrence of the nature referred to in sub-section (1) shall, as early as practicable, cause such remedial measures to be taken as are necessary to prevent or mitigate the environmental pollution.

(3) The expenses, if any, incurred by any authority or agency in respect to the remedial measures referred to in sub-section (2), together with interest (at such reasonable rate as the Government may, by order, fix) from the date when a demand for the expenses is made until it is paid may be recovered by such authority or agency from the person concerned as arrears of land revenue or of public demand.

11.(1) Subject to the provisions of this section, any person empowered by Central Government in this behalf shall have a right to enter, at all reasonable times with such assistance as he considers necessary, any place.

(a) For the purpose of performing any of the functions of the Centre entrusted to him.

(b) For the purpose of determining whether and if so in what manner any such functions are to be performed or whether any provisions of this Act or the rules made thereunder or any notice, order, direction or authorization served, may be given or granted under this Act is being or has been complied with.

(c) For the purpose of examining and testing any equipment, industrial plant, record, register, document or any other material object or for conducting a search of any building in which he has reason to believe that an offence necessary to prevent or mitigate environmental pollution.

(2) Every person carrying on any industry, operation or process

or handling any hazardous substance shall be bound to render all assistance to the person empowered by the Central Government under sub-section (1) for carrying out the functions under that sub-section if he fails to do so without any reasonable cause or excuse, he shall be guilty of an offence under this Act.

(3) If any person wilfully delays or obstructs any person empowered by the Central Government under sub-section (1) in the performance of his functions he shall be guilty of an offence under this Act.

The provisions of the Code of Criminal Procedure, 1973 in relation to the State of Jammu & Kashmir or any area in which that code is not in force. The provisions of any corresponding law in force in that state or areas shall so far as may be apply to any search or seizure made under the authority of a warrant issued under section 94 of the said code or, as the case may be under the corresponding provision of the said law.

12.(1) The Central Government or any officer empowered by it in this behalf shall have power to take for the purpose of anlaysis samples of air, water, soil or their substance from any factory, premises or other place in such manner as may be prescribed.

(2) The result of any analysis of a sample taken under sub-section (1) shall not be admissible in evidence in any legal proceeding unless the provisions of sub-sections (3) and (4) are complied with.

(3) Subject to the provisions of sub-section (4), the person taking the sample under sub-section (1) shall —

(a) Serve on the occupier or his agent or person in charge of place a notice then and there in such form as may be prescribed of his intention to have it so analysed.

(b) In the presence of the occupant or his agent or person.

(c) Cause the sample to be placed in a container or containers which shall be marked and sealed and shall also be signed both by the person taking the sample and the occupier or his agent or other person.

(d) Send without delay, the container or the containers to laboratory established or recognized by the Central Government under section 12.

(4) When a sample is taken for analysis under sub-section (1) and the person taking the sample serves on the occupier or his agent or person a notice under clause (a) of sub-section (3), then,

(a) In a case where the occupier, his agent or person wilfully absent himself, the person taking the sample shall collect the sample for anlaysis to be placed in a container or containers which shall be marked and sealed and shall also be signed by the person taking the sample, and

(b) In case where the occupier or his agent or person present at the time of taking the sample refuses to sign the marked and sealed.

These acts as well as allied rules if enacted along with the following recommendation major pollution problems can be omitted.

Environmental Guidelines for Industries

In India to help minimize environmental pollution and the effects caused by the pollutants the following environmental guidelines are recommended and for siting of industries to ensure optimum use of natural are man-made resources in sustainable manner with minimal depletion, degradation and/or destruction of environment. They are in addition to those directions that are already in existence under the Industries (Development and Regulation) Act.

Areas to be Avoided

In siting industries, care should be taken to minimize the adverse impact of the industries on the immediate neighbourhood as well as distant places. Some of the natural life sustaining systems and some specific land uses are sensitive to industrial impacts because of the nature and extent of fragility. With a view to protecting such areas, it shall maintain distances from the areas listed.

Ecologically and/or otherwise sensitive areas include:

(i) Religious and Historic places,
(ii) Archeological monuments (e.g., identified zone around Taj Mahal),
(iii) Scenic area,
(iv) Hill resorts,
(v) Beach resorts,
(vi) Health resorts,
(vii) Coastal areas rich in corals, mangrooves, breeding grounds of specific species,
(viii) Estuaries rich in mangrooves, breeding grounds of specific species,
(ix) Gulf areas,
(x) Biosphere reserves, grounds of specific species,
(xi) National park and Sanctuaries rich in mangroves, breeding grounds of specific species,
(xii) Natural lakes, swamps,
(xiii) Seismic zones,
(xiv) Tribal settlements,
(xv) Areas of scientific and geological interest,
(xvi) Defense installation, specially those of security importance and sensitive to pollution,
(xvii) Border areas (international),
(xviii) Air Ports, and
(xix) Marine areas.

Pre-requisite

State and Central Governments are required to identify such areas on a priority basis.

Siting Criteria

Economic and social factors recognized and assessed while siting an industry. Environmental factors must be taken into consideration in industrial siting. Industries are therefore required to be sited as well as a striking balance between economic and environmental considerations is required to be maintained. In such a selected site, the following factors must be recognized :

- No forest land shall be converted into non-forest activity for the sustenance of the industry.
- No prime agricultural land shall be converted into industrial site.
- Within the acquired site the industry must be located itself at the lowest location to remain obscured from general sight.
- The green belt between two adjoining large scale industries shall be one kilometre.
- Land acquired shall be sufficiently large to provide space for appropriate treatment of wastewater still left for treatment after maximum possible reuse and recycle. Treated wastewater shall be used to raise green belt and to create waterbody for aesthetics, recreation and if possible, for aquaculture. The green belt shall be half kilometre wide around the battery limit of the industries. For industry having odour problem it shall be one kilometre wide.
- Enough space should be provided for storage of solid wastes so that these could be available for possible reuse.
- Layout and form the industry that may come up in the area must confirm to the landscape of the area without affecting the scenic features of that place.
- Associated township of the industry must be created at a space having physiographic barrier between the industry and the township.
- Each industry is required to maintain three ambient air quality measuring stations within 120-degree angle between stations.

ENVIRONMENTAL IMPACT ASSESSMENT

The purpose of Environmental Impact Assessment (EIA) is to identify and evaluate the potential (beneficial and adverse) of development projects on the environmental system. It is a useful aid for decision-making based on understanding of the environmental implications including social, cultural and aesthetic concerns which could be integrated with the analysis of the project costs and benefits. This exercise should be undertaken early enough in the planning stage of projects for selection of environmentally compatible sites, process technologies and such other environmental safeguards.

While all industrial projects may have some environmental impacts all of them may not be significant enough to warrant elaborate assessment procedures. The need for such exercises will have to be decided after initial evaluation of the possible implications of a particular projects and its location. The projects, which could be the candidates for detailed environmental impact assessment, include the followings:

(i) Those, which can significantly alter the landscape, land use pattern and lead to concentration of working and services population.

(ii) Those which need upstream development activity like assured mineral and forest products supply of down stream industrial process development.

(iii) Those involving manufacture, handling and use of hazardous materials.

(iv) Those, which are sited near ecologically sensitive areas, urban centres, hill resorts, places of scientific and religious importance.

(v) Industrial estates with constituent units of various types, which could cumulatively cause significant environmental damage.

The environmental impact assessment (EIA) should be prepared on the basis of the existing background pollution levels *vis-à-vis* contributions of pollutants from the proposed plant.

The EAI should address to some of the basic factors listed below:

(a) Meteorology and air quality: Ambient levels of pollutants such as sulphur dioxide, oxides of nitrogen, carbon monoxide, suspended particulate matters, should be determined at the centre and at three other locations on a radius of 10 km. with 120 degrees angle between stations. Additional contribution of pollutants and the locations are required to be predicted after taking into account the emission rates of the pollutants from the stacks of the proposed plants, under different meteorological conditions prevailing in the area.
(b) Hydrology and water quality.
(c) Site and its surroundings.
(d) Occupational safety and health.
(e) Details of the treatment and disposal of effluents (liquid, air and solid) and the methods of alternative uses.
(f) Transportation of raw material and details of material handling.
(g) Impact of sensitive targets.
(h) Control equipment and measures proposed to be adopted.

The plans should indicate the details as to how various measures have been or are proposed to be taken including cost components as may be required. Cost of measures for environmental safeguards should be treated as an integral component of the project cost and environmental aspects should be taken into account at various stages of the projects.

(i) *Conceptualization*: Preliminary environmental assessment.
(ii) *Planning*: Detailed studies of environmental impacts and design of safeguards.
(iii) *Execution*: Implementation of environmental safety measures.

(iv) *Operation*: Monitoring of effectiveness of built-in safeguards.

The management's plans should be necessarily based on consideration of resource of conservation and pollution abatement, some of which are enumerated as under:

Liquid Effluents

(i) Effluents from the industrial plants should be treated well to the standards as prescribed by the Central/State Government Water Pollution Control Boards.
(ii) Soil permeability studies should be made prior to effluents being discharge into holding tanks or impoundments and steps taken to prevent percolation and groundwater contamination.
(iii) Special precautions should be taken regarding flight patterns of birds in the area. Effluents containing toxic compounds, oil and grease have been known to cause extensive death of migratory birds. Location of plants should be prohibited in such type of sensitive areas.
(iv) Deep well burial of toxic effluents should not be resorted to as it can result in re-surfacing and ground water contamination. Re-surfacing has been known to cause extensive damage to crop and livestocks.
(v) In all cases, efforts should be made for reuse of water and its conservation.

Air Pollution

(i) The emission levels of pollutions from different stacks should conform to the pollution control standards prescribed by central or state boards.
(ii) Adequate control equipment should be installed for minimizing the emission of pollutants from the various stacks.

(iii) In-plant control measures should be taken to contain the fugitive emissions.

(iv) Infrastructural facilities should be provided for monitoring the stack emissions and measuring the ambient air quality.

(v) Proper stack height as prescribed by the Central/State Government Pollution Control Boards should be provided for better dispersion of pollutants over a wider area to minimize the effect of pollution.

(vi) Community buildings and townships should be built upwind of plant with one half to one-kilometre green belt in addition to physiographical barrier.

Solid Waste

(i) The site for waste disposal should be checked to verify permeability so that no contaminants percolate into the groundwater or river/lake.

(ii) Waste disposal areas should be planned down wide of villages and townships.

(iii) Reactive materials should be disposed of by immobilizing the reactive materials with suitable additives.

(iv) The pattern of filing disposal site should be planned to create better landscape and be appropriate agency and the appropriate, pretreated solid wastes should be disposed according to the approved plan.

(v) Intensive programmes of free plantation on disposal areas should be undertaken.

Noise and Vibration

Adequate measures should be taken for control of noise and vibrations in the industries.

Occupational Safety and Health

Proper precautionary measures for adopting occupational safety and health standards should be taken.

Preventive Maintenance and Operation of Environmental Control Systems

(i) Adequate safety precautions should be taken during preventive maintenance and shut down of the control systems.
(ii) A system of interlooking with the production equipment should be implemented where highly toxic compounds are involved.

House Keeping

Proper housekeeping and cleanliness should be maintained both inside and outside the industry.

Human Settlements

(i) Residential colonies should be located away from the solid and liquid waste dumping areas. Meteorological and environmental conditions should be studied properly before selecting the site for residential areas in order to avoid air pollution problems.
(ii) Persons who are displaced or have lost agricultural lands as a result of locating the industries in the area should be properly rehabilitated.

Transport Systems

(i) Proper parking places should be provided for the trucks and other vehicles by the industries to avoid any congestion or blocking of roads.
(ii) Siting of industries on the highways should be avoided as it may add to more road accidents because of substantial increase in the movements of heavy vehicles and unauthorized shops and settlements coming up around the industrial complex.

(iii) Spillage of chemicals/substances on roads inside the plant may lead to accidents. Proper road safety signs both inside and outside the plant should be displayed for avoiding road accidents.

Recovery-reuse of Waste Products

Efforts should be made to recycle or recover the waste materials to the extent possible. The treated liquid effluents can conveniently and safely used for irrigation of lands, plants and fields for growing non-edible corps.

Vegetal Cover

Industries should plant trees and ensure vegetal cover in their premises. This is particularly advisable for those industries having more than ten acres of land.

Disaster Planning

Proper disaster planning should be done to meet any emergency situation arising due to gas etc. fire-fighting equipment and other safety appliances should be wept ready for use during disaster/ emergency situation including natural calamities like earthquake/ flood.

Environmental Management Cell

Each industry should identify within its set up a department/section or cell with trained personnel to take up the model responsibility required for planning and implementation of the projects.

4

ADOPTION OF GREEN MARKETING PRACTICES BY COOPERATIVE MARKETING AND PROCESSING ORGANIZATIONS - A NEW MILLENNIUM APPROACH

Samwel Kakuko Lopoyetum

Introduction: An Overview

Cooperative Marketing and Processing Organizations and Enterprises are deeling with large number of Agricultural Commodities and Products. The Agricultural Cooperative Marketing and Processing Enterprises assumes greater significance in the New Millennium. They must meet the requirements of the WTO regime and must polish their working operations planning, functions, objectives, enhancement of the scope of Administration and Management to meet the Environmental requirements of producing and marketing their products and services. They must improve their marketing efficiency and effectiveness and incorporate Green Marketing System. The positive side of Indian Agricultural Marketing and Processing System is its perspective competitiveness and mass consumption of Agricultural Commodities and Products. There are higher chances and opportunities to compete in the Global Market Economy and even good chances of expanding exportation in the future. Therefore, they must produce and market environmentally acceptable goods and services. Their Products and Services must be Eco-friendly in all its dimensions and components. The Agricultural Cooperative Marketing and Processing Enterprises must be equipped with proper credit (financial) facilities, adopt proper marketing strategies and

Green Marketing Practices, setup proper storage system and warehousing facilities, adopt Eco-Friendly Marketing and Green Marketing, adopt efficient and effective Eco-friendly processing system, obtain eco-friendly export support system, adoption of the latest technology and modernization to gain higher competitive advantages, use of effective channel of distributions (within and outside the country), introduction of integrated ways and means of handling agricultural commodities and products to attain environmentally acceptable standards, adoption of the best methods and practices of Green Marketing etc.

The cooperative business environment has to adjust the mindset of the existing staff to match their thinking with the new techniques and contents of the job for enhancing Green Productivity. Human Resource Development means all the planned information, education, training, modernization and manpower development activities undertaken by cooperatives so as to create economically effecient organization, capable of providing services by their members.

In fact Human Resource Development has to be conceived as a broad concept which encompass within its ambit not only employees but also members, office bearers, board of directors of the cooperatives. Human Resource Development necessary to acquire and sharpen human capabilities required to perform various functions associated with their present or expected future role. Human Resource Development is the processing increasing the knowledge, the skills and the capacities of all the people in the society.

The Cooperative Business environment has to be Environmental friendly. Cooperative Sector must act and work like Bees which remove the nectar from the flowers without damaging the petals of the flowers. The cooperative sector must go for Eco-friendly development project, to develop the cooperatives in an environmental friendly by manner. The cooperative sector therefore should work out the matching developmental plans which can encourge proper utilization of staff without sacrificing effeciency and profitability and use the natural resources to produce environmentally acceptable products and services.

The Cooperative National Centre for Cooperative Education (NCCE) is playing an effective role in promoting Human Resource Development through its leadership development programme. It has been playing an effective role in sensitizing the cooperative leaders on various issues of concern to the cooperative sector. The aim of leadership development programme is to develop entrepreneurial qualities and to promote professionalism among the cooperative management. The goal before cooperative is to equip the management and workforce with latest management skills to enable them to face the challenges emerging from new market environment with grit and determination. In this way, they can be offered sufficient education and training on the best ways of protecting the Environment, Producing Commodities and Products which are Eco-friendly, reduction in the industrial and use of the best techniques in the Environmental management, they must undertake full-fleged Environmental impact assessment before starting up any cooperative projects. Cooperative industries and manufacturing cooperatives must make use of cleaner technologies for industrial production. They must adopt ISO-14001 and pollution prevention measures in their area of operations to show concern in the Environmental care and Management. The Cooperative Industries and manufacturing units can opt and adopt the Eco-technology for pollution control and environmental management for attaining the sustainable economic growth and development of rural and urban areas.

The National Cooperative Union of India is to build up a strong and knowledge intensive human resources base in Cooperative sector to facilitate environmentally sustainable development of projects. Diversification in the competitive environment is gaining momentum, therefore, cooperative sector must neccessrily create an enlightened, dynamic and forward looking leadership along with the responsive membership to promote Eco-friendly products and protection of environment. The NCUI has to inprove the cooperative sector in aiming at attainment of these goals through various cooperative education (to members) and training (to employees) programme thereby increasing Green Productivity in Cooperative sector through effective and efficient Education and Training Programmes. Some of the programmes run by the NCUI are (a) Cooperative Education and Leadership Development Programmes, (b) Cooperative Training Programmes, (c) Education Field Projects in Cooperatively Under–Developed States, and (d) Cooperative Education Programmes for Empowerment of Women and Youth. The apex level organizations such as NAFED, TANFED, TRIFED, NCDC, NCUI, etc. must promote and support GREEN MARKETING in the Cooperative Sector. Furthermore, an apex marketing organization such as NAFED has exported Agricultural Commodities through well knit and established cooperative marketing system and network (i.e. Marketing of Agricultural Commodities such as Onions, Potatoes, Nigerseed, Sunflower seed, Turmeric and Cashewnuts) valued at Rs.1704 crore have been marketed during 2000-01. It can encourage the production and processing of Eco-friendly and Environmentally acceptable products and services. In the New Millennium, almost all the successful Economic Organizations have changed their working business style and approach to consumerism in the recent times. Many successful and competitive organizations can only produce the products and services, plan, execute and direct all their business activities to maximum satisfaction of the consumers and customers. Nowadays, consumers needs and wants are directly linked to Eco-friendly and Environmentally acceptable products. Therefore, Cooperative Sector can become a leader and model in the *Green Marketing System* in the New Millennium. It is highly imperative to promote organizational link-up system to encourage growth and development of *Green Marketing Cooperartive Organisations and Enterprises in the New Millnnium.* The following link-up system can be encouraged, viz., (i) Linking Green Marketing Cooperative Enterprises with credit (financial) organizations, (ii) Linking Green Marketing Cooperative Enterprises with processing organizations. (iii) Linking Green Marketing Cooperative Enterprises with processing and Export- oriented organizations, (iv) Linking Green Marketing Cooperative Organizations with Trading and Production System, (v) Linking Green Marketing Cooperative Enterprises with inter-Cooperative Oraganizations (within and outside India).

Environmental Management and Implementation issues are directly and indirectly affecting and influencing all human and business organizational activities. The entire society and community have become sensitive and concerned with the natural environmental and depletion of natural resources. The *New Concern and Approach to Marketing of Green Products* must be adopted by the Cooperative Marketing and Processing Organization in the New Millennium. Here, Green Marketing refers to marketing of products, commodities and services considered. Envi-ronmentally friend (Eco-friendly) and by doing so, they make Marketers to be Environmentally responsible directly and accountable in the long-run. The demand for *Green Marketing of Products and Services* have been necessitated by the Expanding demand by the consumers (both in developed and developing countries). Even many Governments across the developed and developing countries are increasingly becoming more concerned about the promotion of *"Green Marketing System and Business Activities"*. A large number of rules, procedures, acts, laws, guidelines, controls and regulations have been framed and instituted by various governments to promote *"Green Marketing."*

Consequently, Cooperative Processing and Marketing Organizations and Enterprises must adopt all the Green Marketing Practices and Methods in order to protect Environment and make it more sustainable in the long-run. This particular paper has highlighted and described the Green Marketing problems and prospects of adopting the Green Marketing. New approach to be adopted by Cooperative Processing and Marketing Organizations, Adoption of New Marketing Strategies to ensure effective and efficent Green Marketing System in Cooperative Organizations and Enterprises in the New Millennium.

Dimensions of Agricultural Cooperative Marketing and Processing Organizations

The Agricultural Cooperative Marketing and Processing Organizations assume great significance in the era of economic liberalization and globalization because the current open and free system have percolated into all the areas of markets and in all areas of activities touching agriculture and allied sectors that have been fully liberalized. Therefore, there is an urgent need to make agricultural cooperative marketing and processing organizations to function more effectively and efficiently and produce eco-friendly products. The acceptance of WTO rules and regulations by the Government have created more competition between India and many overseas countries, therefore Indian Cooperative Marketing System and Processing Organizations spared. It is very difficult for agricultural marketing and processing cooperative organizations to meet the requirements of the WTO without polishing their working operations, functions, objectives, scope administration, management and thorough improvenent of the over all management of the marketing efficiency.

In the era of economic liberalization (EL), privatization (P), globalization (G) and WTO regime, it is essential to formulate suitable mechanism by the Government to protect the domestic market for the goods and products of the farmers. Particularly the Green Products and Eco-friendly Commodities. The agricultural marketing and processing cooperative organizations must put up concerted efforts to develop market intelligence, Effective Market Information System (EMIS) and market research to update their Data Bank Management System (DBMS) and explore the actual export potentialities of the agricultural commodities. This is highly relevant for the Green Products and Protection of Environment. Some support measures such as subsidy for research and development in the Green Cooperative Marketing field may be indispensable in this new era of open market and highly essential to protect the agricultural marketing sector and its domestic market, particularly those declining with Green Productivity. The positive side is that, Indian agricultural marketing system is competitive due to its mass consumption and bright chance of exportation in the future. The contribution of agricultural sector in the Indian economy amount to 27 per cent of GDP and the livelihood maintenance of 70 per cent of the population, therefore, the agricultural cooperative marketing societies and processing organizations can transform the socio-economic conditions of many rural folks and the entire rural economy.

The world-wide phenomenon of Economic Liberalization (EL) and Globalization (G) have created highly competitive milieu for which they must create and develop highly competitive Green Products brands, suitable green labels and produce highly competitive green products, improve and upgrade their quality, reduction in cost of production, use of cost effective techniques and methods, acclimatize the farming community and small/ medium enterprises with the provisions of the green marketing and productivity. Also, the WTO agreement on agricultural cooperative and provision of greater market access for agro-processed products which are having environmental friendly approaches. The agricultural cooperative marketing must be well equipped with proper credit facilities, proper marketing strategies, proper storage facilities, efficient and effective processing methods, export support, adoption of latest technology to gain competitive advantages, effective channel of distribution, introduction of the integrated ways and means of handling agricultural commodities by agricultural cooperative marketing organization.

Most of the agricultural farming commodities are produced by the small and marginal farmers (78 per cent) that depend mostly on agricultural cooperative marketing societies and processing organizations for disposing of their produce and commodities. They must be made aware of the challenges, prospects, threats and opportunities of Economic Liberalization (EL) and Globalization. The small and marginal farmers are like to face many disadvantages in the era of economic reforms mainly due to size of their lands, fragmentation, dispersal of their holdings, spatial area for selling are limited, inadequate transport facilities, the limited amount of produce, lack of uniform quality, inadequate financial resources immediate selling of their commodities after harvest and at extremely low prices. Therefore, they must be encouraged from time to time to use agricultural cooperative marketing societies to solve their impinging problems. In this era of Globalization of market environment, they must set up farmer-oriented cooperatives, farmer-controlled cooperatives, professionally managed agricultural cooperative, profitable trade and viable business, for this to happen, they must organize themselves into very strong cooperative organizations or self-help group (SHGs) which have green productivity in their minds and approaches. Even, the regulated markets under the control of agricultural produces market commitee existing in various states, must be promoted and created to be better or even greater system must be initiated and developed to provide better infrastructural facilities.

The process of liberalization has been increasing and accelerated by the introduction of New Economic Reforms, as such the cooperative sector has to newly buildup a strong agricultural marketing system and processing laying more emphasis on cost-effectiveness and cost competitiveness, production of high quality standard of products, assimilation of modern management practices, adoption of information technology, promotion of Green Productivity, Protection of Environment, Production of Eco-friendly Products and Services and constantly encouraging exports of green products. For instance, NAFED has exported agricultural commodities through established cooperative marketing system i.e., onions, potato, nigerseed, sunflower seed, turmeric and cashewnut valued at Rs. 1704 crores during 2000-2001. The success and growth of agricultural cooperative marketing societies would certainly depend on the members loyalty and economic participation of all members based on commercial benefits, which again depend on the harmonious co-existence/relationship between members, paid employees, board of directors (BOD), elected board of management and the public at large. Moreover, if cooperative marketing and processing organizations need to succeed, they must have efficient and effective cooperative leadership, cooperative linkages, training members, education and awareness, honesty and democratic management. All these factors are very essential for the positive growth and success of agricultural cooperative marketing organizations which are dealing with green productivity in the era of economic liberalization (EL) and Globalization.

Relevance of Agricultural Cooperative Marketing Management under Economic Liberalization and Globalization

Cooperative marketing system remains to be even more crucial in the era of ecnomic liberalization and globalization, particularly, in the areas of production, finance, purchase, management, personnel, sales, processing, methods, systems etc., which must create time, place and possession of utilities of goods and services produced by the farming community. It is highly essential in this era that, the commodities reach its destination (consumers) all the right time, right place, right quality and quantity and offered at the right price. In this era, it requires a network of effective and efficient agricultural cooperative marketing organization which would in turn provide remunerative prices to the farmers (producers). At the same time, farmers must procure prompt services from their own cooperative marketing society and ensure sustainability of the marketing system. There is a need to remove unwanted intermediaries/agent and exploitative marketing systems by producing proper good services and assuring reasonable returns to the producers and must provide full satisfaction to the customers. In India, agricultural cooperative marketing system plays a major role in the agricultural marketing system. For many years, cooperative marketing system has ensured remunerative returns to the agricultural producers and exerts enormous influence on the market behavioural system. Therefore, it must be made more effective and efficient in this era of economic liberalization and globalization competition.

Imperative Need for Efficient and Effective Agricultural Cooperative Marketing and Processing System in the Era of Liberalization and Globalization

In the new economic era, agricultural cooperative marketing operations and production systems have been commercialized leading to commercialization of agriculture by using scientific methods of agricultural operations and advance production methods. Green Productivity is needed in this era of commercialization. Farmers in developed and developing countries must need certainly higher income and full-time employment in agricultural sector. Therefore, there is an urgent need for efficient and effective sale of agricultural produce and commodities which are highly eco-friendly and environmentally focused in its approach.

They must have better chances of dominating market exploitative nature of the middlemen, intermediaries and often undesirable marketing functionaries, must be removed with the formation of highly competitive marketing and processing cooperative societies. Thus, they must intervene between the agricultural farmers and the ultimate consumers, by ensuring proper satisfaction (total) to the consumers and proper margin to the producers. Similarly, they must remove all the malpractices such as the unnecessary and arbitrary reduction of the prices and encouraging prompt payment or regular payment to the producers. They are ultimately destined to reduce the price spread between the producers and consumers in the long-run. They must provide better returns to agricultural producers. They must also provide large scale loaning (credit) facilities and remove unnecessary exploitation of cooperative credit system. They must provide good (high) quality of green products and quantity of goods and services. They must remove and eradicate frequent manipulations of weight and scale. They must remove collusion between broker and buyers. They must curtail fluctuations of prices and even hold the price line. They must remove unnecessary form of speculations and price manipulations. It can be an agent for recovery of agricultural loan and credit issued by the CCBs and PACs (Linking of credit with marketing). In total, it can form as a giant fulcrum for having all the essential features of integrated programmes of cooperative development and implement all the activities for the total socio-economic development of the rural (agricultural) farmers and rural economic development.

Uniqueness of Cooperative Marketing and Processing System in the Era of Economic Liberalization and Globalization

If the cooperative marketing organizational structures are made effective and efficient, the cooperative marketing system can fulfil its objectives, address the concern of the agricultural produces, provide adequate market power to the producers, face competition with higher devotion, larger market coverage and determination to fulfil the objectives capture a large share of market share, etc. In the era of economic liberalization and globalization, there is a greater need for effective and efficient cooperative marketing leadership and total menber commitment in the operations of the cooperative processing and marketing system. The internal economics of cooperative marketing system would enhance the vertical integration of the cooperative movement and improve the functional relationship of the member producer in the long run. It is highly essential for the agricultural cooperative marketing organization to produce goods and services where their marginal cost (MC) equalled to the Marginal Revenue (MR) (assuming the lacunae of this concept). The cooperative marketing organizations must aim at optimization objective and dispensation of maximum benefits to the members of the marketing organization through equitable distribution of earnings and plough back to the members the benefit in proportion to their patronage volume or patronage dividend. The cooperative marketing organization must acquire internal and external stability accompanied by ownership and control of its management affairs. This can be achieved when there is complete loyalty and cohesiveness in the membership structure and organization of the cooperative marketing and processing system. The paramount uniqueness of the cooperative marketing organizations is its ability to perform multi-functional and multi-facet roles in sorting out the major problems of agricultural farmers and encourage green productivity. It can fulfil the socio-economic-cultural-psychological dimensions of the agricultural producers and community at large.

Value-added Cooperative Approach

The value-added cooperative approach is essential in the cooperative processing and marketing system. to face the competitive and open market economic system prevailing in the country at the moment. The quality of products and services must be enhanced to meet the prevailing market demand and wants of the consumers such as the green products and green services. In the globalized market, the author has advocated for dynamic renaissance in the entire agricultural cooperative marketing system and processing organizations with focus on value addition and quality upgradation by linking cooperative marketing organizations with efficient and effective value additions cooperative organizations. Thus, eventually would generate additional benefits to the owner-users of the cooperative marketing societies and processing organizations.

The new generation cooperative marketing and processing societies must be oriented towards providing comparable products, services and prices to their competitive counterparts such as the private organizations, companies, MCNs, corpotations etc. They must dominate in the marketing and distribution channels, processing and marketing organizations must adopt better distribution strategies to face emerging competitive global markets in the new millennium. For this concept to succeed, they need dynamic link-up organizational systems in this era of open competition and economic liberalization and globalization.

The organization link-up system is imperative in cooperative marketing system to become viable and profitable. The essential link-up systems are:

1. Linking of cooperative marketing system with credit organizations/financial institutions.
2. Linking of cooperative marketing system with processing organizations.
3. Linking of cooperative marketing system with processing and exports oriented organizations.
4. Linking of cooperative marketing system with Green Production System/Green Productivity Organizations.
5. Linking of coopertative marketing system with consumer cooperative organisations which are pro-green productivity.
6. Linking of cooperative marketing system with housing cooperative organizations which uses environmentally friendly structures.

7. Linking of cooperative marketing system with inter-cooperative banking organizations which supports green productivity.

Application and Adoption of Web Marketing Practices in all large sized Agricultural Cooperative Marketing and Processing Societies in the Era of New Economic Liberalization and Globalization

The Agricultural Cooperative Marketing Societies and Processing Organizations must recognizing that, there is no single, foolproof strategy for gaining competitive advantage other than by using the internet. Every agricultural cooperative needs to work through a process of assessing its environment, identifying opportunities, overcoming barriers to electronic commerce and designing and implementing online programs that will add value to both the organizations and customers. Customer satisfaction, through customer long-term profitability being the fundamental wheels on which the marketing objectives and strategy could rest in and drive towards growth and profitability, the significance of web marketing cannot be undermined, that too in the era of economic liberalization and globalization, web marketing practices uses almost all kinds of technological savvy, media and communication networks such as the internet, use of personal computers, etc. Some of the web-marketing and e-commerce practices includes the following:

The use of personal computers, modems, electronic mail and bulletin boards, even facsimile machines geve organizations new media for reaching potential customers who are the users eco-friendly products, eco-technology and environmental-friendly products and services. Now even the smallest organization can maintain computer data bases of customers names to integrate their marketing campaigns of the green products and green productivity. But what is coming is even more dynamic—the global information highway, and with it an interactive revolution. Advertising is evolving into two way mediun in which consumers with PCs, modems, CD-ROMS and TV can choose what information to access and then spend 20 minutes or more drilling down to the indepth product information they desire inluding the information on the green products and green productivity.

Use of Digital Interactive Media: The advent of the information highway has brought a new form of media (electronic) system Digital Interactive Media are channels of communication with which the audience can participate actively and instantly. With a computer keyboard and an electronic card reader booked up to their TV, audiences can now purchase products via cable TV. They can vote or participate in management shows and even give orders of their green products and services.

> *Use of Internet:* The internet, a separate system is the world's largest computer network, working through the online services. Unlike the PC online system, here the entire gamut of advertisers can be easily accessed and can navigate their way around the world from one web site to another, in this way, the customers and consumers of the green products and promoters of Green Productivity can access all the details and information they require from various organizations and agencies.

Adoption and Adoption web marketing tools by Cooperative Marketing and Processing Organization in the era of economic liberalization and globalization

The Era of globalization is composed of dynamic aspects and brought enormous sea changes and great opportunism with respect to technological developments. Marketing being the lifeline of any business organization has to be standardized and have a global outlook. Cooperative Organizations and allied organizations are enlarging internationally: even small business organizations today have an international perspective. The small and medium scale enterprises and other rural based industries have superiority shown their outstanding performance in the growth of many developing nations such as India and Kenya. Green Productivity in small and medium scale enterprises are highly imperative and essential in the New Economic Environment. Thus, it becomes inevitable and imperative for agricultural cooperative marketing society and processing organizations to adopt high technology in almost all areas of management operations. But it is essential to have thorough evaluation of agricultural cooperative marketing and processing organizations viability and profitability before signing in for any technological upgradation or modernization of business activities.

The agricultural cooperative marketing and processing organizations should not lag behind during the era of technological advancement which is taking place to enhance the transformation process in the entire economy. They should keep changing, develop core competencies, develop strategy to tackle the situation and adopt various marketing skills to face the world of heavy competition. It is better for the cooperative organizations to know that electronic marketing has given for unbelievable challenging and transformational cases. Since the electronic marketing encompasses all the activities such as electronic selling, electronic CRM practices, electronic advertisements, electronic promotional activities etc. The agricultural cooperative marketing and processing organizations are in no way exempted in this competitive world. The following are few suggestions/recommendations which can enhance the adoption and adaption of the web marketing system of Green Products and promotion of the Green Productivity.

- Fundamentally, the agricultural cooperative marketing and processing organizations to evaluate their strengths and weakness in adopting the web marketing techniques / tools. This clears the row towards better marketing programmes and offers the advantages of the media are assessed so as to reap maximum benefit from online and web marketing of green products.
- Through different schemes and projects and agricultural cooperative marketing society and processing organizations can put forth their cases before the concerned authority in order to get financial and other types of assistance for promoting the Green Productivity approaches.

- When the agricultural cooperative marketing societies are going in for web marketing and other online marketing methods, the quality of the products/services should be kept in mind. There should be no compromise in the quality of the products advertised by these agricultural cooperative marketing societies.
- Proper and dedicated co-ordination of all the functional departments is indeed of utmost necessity as a pre-requisites for adopting the web marketing system for the promotion of green products and the green productivity.
- Highly skilled personnel should be employed for purpose of enhancing the growth prospects of agricultural cooperative marketing and processing societies particularly in marketing of the green products and eco-friendly products.
- Through cost benefit analyses are to be carried out before commencing for web marketing activities for the green products and eco-friendly production of products.
- Computerization and mechanization side by side would prove beneficial for the further development of the atgricultural cooperative marketing enterprises and processing organizations.
- Instead of the narrow approach of agricultural marketing, based on low investment of individual cooperative society, the focus may be shifted to agricultural cooperative marketing society as a dynamic group. Changes could be expected or accrued in the delivery channels for the small borrowers, agriculturists and unorganized sectors of the economy who are declining with green products and green productivity through effective and effecient functioning of agricultural cooperative marketing and processing systems.
- There has to be a change in the work culture and strong work ethics on the part of agricultural cooperative marketing and processing employees.
- The work culture has changed most of the organizations

due to growth of computer network and World Wide Web. Multimedia presentations have become a favourite for conveying ideas, messages, project reports, which are Eco-friendly and Green Project Reports, Green Products information systems, describing marketing, presenting annual growth in business commercial houses of various business organizations.

- The agricultural cooperative marketing society may adopt computerized decision making system that gives the purchasing agent time for a more thorough evaluation of possible vendors in more complex buying situations. Purchasing agents can devote more time to direct negotiation with potential suppliers and assume an expanded role in inter-departmental activities of the cooperative organizations.
- The Agricultural Cooperative Marketing Society and Processing Organization should also have responsive marketing strategies on the web system and be prepared to adopt marketing changes to deal with technological environments. Because the most recent wave of technological changes are as dramatic as any in history, the implications for organizational decision-making are highly profound in displaying their green products and Eco-friendly production system.

Suggestion's for Policy Formulations and Recommendations

The following suggestions are offered to make agricultural cooperative marketing and processing societies more vibrant, viable, sustainable and profitable in the era of economic liberalization and globalization.

- Finance is the life blood of any business organization, therefore, cooperative organizations are not exceptional. Lack of funds is a crucial problem in many marketing and processing cooperative societies. Permitting each and every agricultural cooperative marketing and processing societies to get credit from SBI, Canera Bank, CCB's, NABARD etc. and other commercial banks without prior approval from Registrar of Cooperative Societies might be the remedy for the financial problems faced by cooperative organizations.

- The paucity of strong storage facilities in cooperative marketing and processing societies could be overcomed if the central and state warehousing corporations are prepared to extend this facility atleast for a temporary period, after which they may be asked to create their own storage facilities in the future.
- They may be encouraged to appoint trained and professional employees therefore, the performance relating to processing classification, standardization and grading of agricultural produces could be ensured by well trained and professionlised staff. The officials of the society can be encouraged to improve their business ability and enhance the cooperative societies image in the open market economy.
- It is essential to create awareness through educating and training through extension system of the agricultural farmers about the benefits obtained from the marketing and processing cooperative organizations. Also, the ways, means, rules, regulations and procedures of agricultural marketing operation and processing are informed and made useful to all the farmers. They should market their produce through cooperative system and thereby to reap the direct and indirect benefits from proper use of the marketing and processing organizations.
- The electcd Directors/Board of Directors (BOD) must be given training, Extension Services, and adequate so as to enable them to play a vital role in deliberations on the policy decision-making process and effective administrations of agricultural marketing and processing cooperatives. They must understand the advantages and benefits of Green Products and Green Productivity.

- Arrangements can be made to obtain adequate expertise and technical knowledge on the management and administration of agricultural marketing and processing cooperative society which are dealing with Green Productivity.
- Proper communication and transportation system must be implemented in Rural Areas for the benefits of the agricultural farmers.
- The problems of inadequate and underdeveloped state of transportation system must be addressed immediately by announcing Rural Access Road Authority of India (RARAI) to ensure that, the rural feeder roads are made all weather roads even to the Hilly areas and regions of the country.
- The loyalty and active participation of the members, Board of Directors, staff and public must be frequently solicited through techniques such as Customer Relationship Management (CRM), Public Relationship Management (PRM) etc.
- The National Cooperative Development Corporation (NCDC) must enhance, the quantity and quality of the loan and financial assistance given to the agricultural marketing cooperative society, they must reduce their rates of interest (even if they be given benefits of 0.8 per cent as compared to other financial institutions). It must increase the activity and mode of release of funds for *Revitalization and Strengthening* of the share capital of marketing and processing societies. Emphasis must be laid on the Green Products and Green Productivity.
- Incentive schemes and programmes may be initiated by the Agricultural Marketing Society to provide incentives to the progressive farmers, particularly, those who have patronize fully in the business affairs of cooperative societies and engaged themselves in the production of the green procucts and promotion of green productivity.
- All agricultural marketing processing societies, must be

made viable and profitable duly by honest and dedicated staff of the cooperative society. Concerted efforts must be made to encourage the most honest and dedicated staff members by giving them *awards* and *rewards* for their good work. Even the BOD and committee members may be considered for rewards for their good performance, particularly enhancement and promotion of the Green Products (Eco-friendly products and Green Productivity Systems).

- They must learn how effectively and efficiently to compete more meaningfully in the open market with all other types of green based organizations within the country and outside the country.
- Adequate measures must be taken up to revitalize and strengthen weaker agricultural marketing and processing cooperative societies. A suitable scheme may be initiated and launched to revive weak or defunct agricultural cooperative marketing and processing societies.
- They must be quality-oriented and produce products with added value to earn better prices for producers and satisfy consumers with green products.
- They must undertake proper Research and Development (R&D) in various aspects of agricultural cooperative activities. Particularly, they must deal with the green products research and development.
- As a matter of policy issue, the adoption of Information and Communication Technology (ICT) in the agricultural marketing and processing cooperative societies must be made mandatory in the era of EL&G.
- Adoption of both Direct and Indirect marketing systems can boost the sales performance of the societies. They can directly supply their green produce to institutions such as government institutions, colleges, companies etc.
- They must undertake marketing extension and training programmes to enlighten and brighten farmer on the negative and positive effects of open market economic

system and benefits of the green products and green productivity. Also on the best methods of marketing their green produce and overcome the cut throat competition caused by the introduction of new economic reforms in the country.

Conclusion

To sum up the agricultural marketing cooperative society and processing organization capture higher share in the open market system, through production of Eco-friendly and green products. They can reap and gain enormous benefits and harvest the fruit of their hard work. If they sow the right seeds at the right place, at the right time. This is the time to promote and encourage the Green Products and Green Productivity ventures by cooperative marketing and processing organizational methods of doing business and transactions to face cut throat competitions in the open market. They must use the latest methods of doing business, such as the use of E -marketing system, web-marketing system, E-communication system and other marketing friendly technological innovations can be adopted by the agricultural marketing and processing cooperative societies to promote green products and green productivity. They must move forward with determination, courage, faith, commitment, integrity, fraternity, loyalty and capture the entire marketing system of green agricultural commodities/products in this particular era of economic liberalization and globalizations.

REFERENCES

Mamoria, C.B. and Joshi, R.I. *Principles and Practices of Marketing in India.* Kitab Mahal, Allahabad.

Hajela,T.N., *Cooperation—Principles, Problems and Practice.* Konark Publishers Pvt. Ltd., Delhi, 1997. pp.437 to 438.

Basavaraja Banakar and Kunnal, I.B., "Agricultural Cooperarive Marketing in India" *The Cooperator*, NCUI, New Delhi, October 1986. Vol. XXIV.

Sethuraman, K., *Agricultural Marketing.* Margham Publication, Chennai, 2000, pp.16 to 17.

Boyd, *et al.* (1998), *Marketing Management* (Third International Edition 1998) Irwin McGraw-Hill Publishing Company, New York.

Bulletin, Sep. 2003 (No.5) Defence Research and Development Organization, Directorate of Defence Research and Development Organization, Ministry of Defence, New Delhi-110054.

Samwel Kakuko Lopoyetum, *et al.* (April 2003) "Fundamental Management Dimonsions of Cooperatives : Inperative Issue", *Indian Cooperative Review,* Vol. 40 No. 4, pp. 239 to 247.

5

GREEN PROCESSING THROUGH NATURAL / VEGETABLE DYEING IN SMALL DYEING UNITS

R. Seerangarajan

At present there is a widespread awareness on the use of Natural/ Vegetable dyed textile materials due to its merits like eco-friendly processing and majority of the vegetable dyes do not pose toxic/ allergic problems to the wearer. Many garment producing units have started using vegetable dyes to colour their yarns and thereby, strengthen their marketing base. One can observe the increased customer preferences over natural/herbal products in respect of costumes and attires. Thus, natural raw material resources have to be identified and harnessed for suitable applications to make competitive advantage in market which may otherwise go as waste. Cultivation and processing of vegetable dye materials can provide employment to millions of rural people also. Thus the use of vegetable dyes could be propagated for three reasons, namely eco-friendly, health friendly, and for providing employment in rural areas. Vegetable dyed textile goods have earned a niche market globally and many textile producers have explored such market potentials in Indian and International arena. Further these plant sources are needed to be identified in the respective areas for extracting suitable colours. Most of the vegetable dye sources are grown as wild herbs/plants and few dye plants are cultivated which are having very high market potential. The present work is undertaken to evaluate the fastness characteristics of the selected vegetable dye sources and its applicability to cotton textiles and garments for enhancing green productivity of the wet processing. Vegetable dyes and its application on cotton textiles not only enhance green productivity and facilitate rural employment generation.

Objectives of the Study

The objectives of the present work are as below:

1. To develop appropriate methods of extraction of dye from the identified new sources, namely Red gram leaves (*Cajanus Cajan*).
2. To test the solubility of some of the selected vegetable dye sources.
3. To generate simplified dyeing methods from the extract for cotton fabrics and garments.
4. To evaluate fastness characteristics of the dyes to external agencies like washing and rubbing.

Importance of the Study

Due to environmental issues and other health hazards, the application of synthetic dyes will be a problem to the community as they pollute the environment and other water bodies. Hence, use of natural dyes will be the alternate solution for saving the nature and to promote rural employment through collection of plants, processing of dye source and dyeing of textile goods. Thus the present study will be more relevant to the present day's environmental issues and promoting rural development and green productivity.

Design of the Work

Materials and Methods

(a) 100 per cent cotton fabric.
(b) Raw material used in the dyeing process.
  (i) Red gram leaves (*Cajanus Cajan*).

All the above said raw material is available in Dindigul district. This natural dye source was dried and powdered.

All the textile materials used in the study were tanned with the solution of myrobolan for enhancing the application of colour.

(c) *Synthetic Mordants*: The study used the following synthetic mordants.

(i) Alum (Aluminium sulphate);
(ii) Copper sulphate;
(iii) Ferrous sulphate; and
(iv) Potassium dichromate.

Methods

Preparation of Fabric

The fabric was scoured and bleached using the standard procedures.

Scouring

Scouring was done with sodium hydroxide.

Bleaching

Bleaching was done with hydrogen peroxide followed by a souring treatment and washing.

Types of Pretreatment

To improve the dye-ability of natural colorants on cotton, the following types of pretreatment was experimented.

Tanning with myrobolan followed by mordanting with any one of the metallic salts.

Tanning with Myrobolan

To extract the tannin from the myrobolan, the myrobolan fruits after removing the seeds are ground to fine powder and the required quantity of it is boiled in water for 1 hour and then filtered.

The bleached fabric is impregnated using the following recipe.

ML Ratio: 1:10
Solution of Myrobolan: 30% (OWF)

Temp : room temp
Time : 30 minutes

Then the fabric is squeezed, dried for 24 hours and dusted to remove any superficially deposited myrobolan particles.

Another set of bleached cotton fabrics were subjected to synthetic mordants as below:

Recipe

ML Ratio:	1:30
Temp :	60-70° C
Time:	30 Minutes
Mordants:	
Alum:	10%
Copper sulphate:	4%
Ferrous sulphate:	3%
Potassium Dichromate:	2%

The tanned fabric without intermediate washing was mordanted as per the parameters mentioned above. The fabric was then rinsed in cold water and dried in shade.

Extraction of Colourants

The dyeing material namely Red gram leaves was boiled in water with ML ratio 1 : 30 for one hour and then filtered to yield colourant extracts.

Dyeing

Recipe:

ML Ratio	:	1: 30
Dye	:	50%; 100%; 150%; (OWF)
Alkali	:	2% - 4%
Temp	:	Boiling
Time	:	1 hour

The tanned and mordanted fabrics were dyed in the bath maintaining the above said parameters. The fabric was then rinsed in cold water and dried in shade.

Analysis of Vegetable Dyed Cotton Fabrics

Analytical framework provides the base for the evaluation of the characteristics of the vegetable dye selected in the present exercise.

As discussed in mordanting and dyeing were done with Red gram plant leaves—Red gram leaves (*Cajanus cajan*)

The dyeing plan for Red gram leaves will be

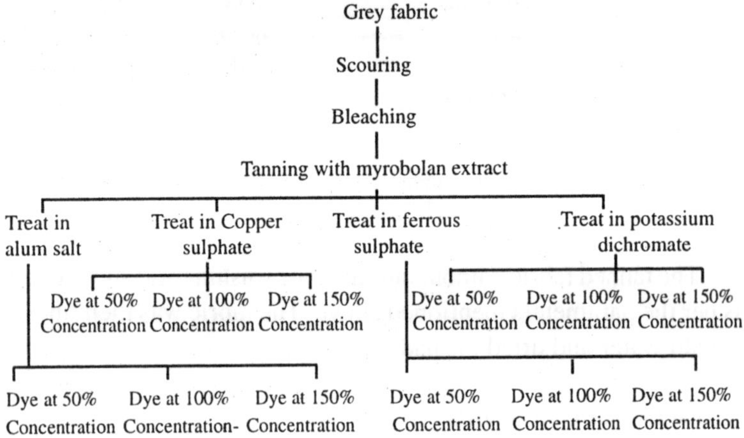

Dyeing with Red gram leaves will have the above scheme of dyeing plan with four metallic mordants.

The dye extracts were prepared based on the weight of the material. 50 per cent concentration of the vegetable dyeing will have the following procedure :

> Let the weight of the fabric to be dyed = 100 gm
> Then 50% of the Red gram leaves will be = 50 gm
> Extract the dye solution by boiling in 30 times water, i.e. with 3000 cc of water.

Dye the goods in prescribed temperature, time and with right chemical additions. After dyeing the material is squeezed, washed and dried in shade.

After thorough drying the goods are evaluated for the following fastness characteristics :

1. Rubbing fastness (IS : 766-1956)

(a) Dry fastness rating ranging from 1-5
(b) Wet fastness rating ranging from 1-5.

2. Washing fastness (IS : 3361-1979)

(a) Change in colour
(b) Staining test.

All the above are evaluated through gray scales.

Red Gram leaves

Red gram leaves are cultivated throughout India and as rain fed short season crops it yields red gram which is a major pulse in Indian foods. After harvesting the red gram plant is used as bio-mass for meeting the energy requirements in rural areas. While doing so, the leaves are dusted off and the dry stalk or stem alone is used as firewood. It has been found through this exercise, that the leaves of the red gram plant after harvesting the red gram could be used as dye source from vegetable origin which otherwise go as waste.

1.1 As done in the previous dyeing the scoured and bleached cotton fabrics were treated with four metallic salts separately as below:

Fabric + Alum + dye at 50%/100%/150% concentrations of red gram leaves.
Fabric + copper sulphate + dye at 50%/100%/150% concentrations of red gram leaves

Fabric + ferrous sulphate + dye at 50%/100%/150% concentrations of red gram leaves.
Fabric + potassium dichromate dye at 50%/100%/150% concentrations of red gram leaves.

1.2 Colour yield with Red Gram leaves

Red gram leaves yield a range of colours from brownish yellow to reddish yellow/ khaki with different metallic salts as below:

(i) Aluminium sulphate with red gram leaves as dye source yields pale brownish yellow to deep bright brownish yellow at concentration ranging from 50 per cent to 150 per cent.
(ii) Redgram leaves with copper sulphate as mordant yield range of colour from pale yellowish brown to dark yellowish brown as concentration of the dye increased from 50 per cent to 150 per cent.
(iii) Ferrous sulphate mordant with red gram leaves dye yielded pale rock brown to deep rock brown as the concentration of the dye increased from 50 per cent to 150 per cent.
(iv) Red gram leaves dye yielded pale reddish khaki to deep reddish khaki in the above stated concentrations. Therefore, red gram leaves may be used as a dye source to dye cotton material to get shades in brownish yellow, yellowish brown, rock brown and reddish khaki.

1.3 Fastness Rating of Red Gram Leaves Dye (Table 5.1)

Rubbing Fastness at Dry Condition

The rubbing fastness of alum mordanted fabric dyed with red gram leaves have demonstrated fastness rating from 4 to 4/5. At a concentration of 150 per cent of the dye, fastness rating was above 4 and 4/5.

Similarly, potassium dichromate mordant has shown better dry rubbing fastness rating ranging from 4/5 to 4. Here, 50 per cent and 150 per cent concentration have shown higher fastness rating of 4/5 than at 100 per cent concentration. Such variation may be due to the reactivity of the mordant in different concentrations. Further evaluation of dye take up tests will aid better findings.

Other two mordants namely, copper sulphate and ferrous sulphate do not exhibit satisfactory fastness rating at 100 per cent and 150 per cent concentrations. Only 50 per cent concentration have yielded fairly better rating of 4.

The overall fastness grading varied from 4/5 to 3.50 per cent and 150 per cent concentration yielded better fastness in dry rubbing condition.

1.4 Rubbing Fastness at Wet Condition

Red gram leaves dye has scored wet rubbing fastness rating ranging from 4 to 1/2. Consistent fastness rating in this test were recorded with copper sulphate mordant.

Sulphate and potassium dichromate mordants yielded moderate fastness ratings and not as good as dry rubbing test. Among the four mordants ferrous sulphate and alum have shown poor fastness rating with wet rubbing test.

From the above, it could be inferred that copper sulphate mordant with the dye has recorded fairly moderate wet rubbing fastness followed by potassium dichromate, alum and lastly the ferrous sulphate mordant. Therefore, Red gram leaves perform moderately with copper sulphate and potassium dichromate mordants for application to apparel products.

1.5 Washing Fastness with Red Gram Leaves

Washing Fastness (Change in Color Test)

As per Table 5.1 the fastness rating in respect of washing—change in colour has shown fairly good fastness rating ranging from 5 to 4/5.

Table 5.1 : Dyeing with Red Gram Leaf

Fastness Properties	*Mordant*	*Alum*			*Copper Sulphate*			*Ferrous Sulphate*			*Potassium Dichromate*			*Overall Grading*		
	Dye %	50	100	150	50	100	150	50	100	150	50	100	150	50%	100%	150%
Rubbing	Dry	4	4	4/5	4/5	2/3	2	4	2	4	4\5	4	4/5	4/5	4	4/5
	Wet	3	3/4	2	3	2/3	3/4	4	2	1/2	1/2	3	3	3	3/4	3
Washing	Change in Colour	5	4/5	4/5	5	4/5	4/5	4/5	4/5	4/5	4/5	r 5	5	5	5	5
	Staining on Cotton	4/5	4	4	3/4	4	4/5	4	3/4	4/5	4/5	4	4	5	4	4
Colour yield		Bright Brownish Yellow	Medium Brownish Yellow	Dark Browhish Yellow	Pale Brownish Yellow	Medium Yellowish	Dark Yellowish Brown	Pale Rock Brown	Medium Rock Brown	Deep Rock Brown	Pale Reddish Khaki	Medium Reddish Khaki	Dark Reddish Khaki			

	50	100	150
Alum	4	4	4.5
Copper sulphate	4.5	2.3	2
Ferrous sulphate	4	2	4
Potassium dichromate	4.5	4	4.5
Overall	4.5	4	4.5

RED GRAM LEAF
RUBBING FASTNESS—DRY CONDITION

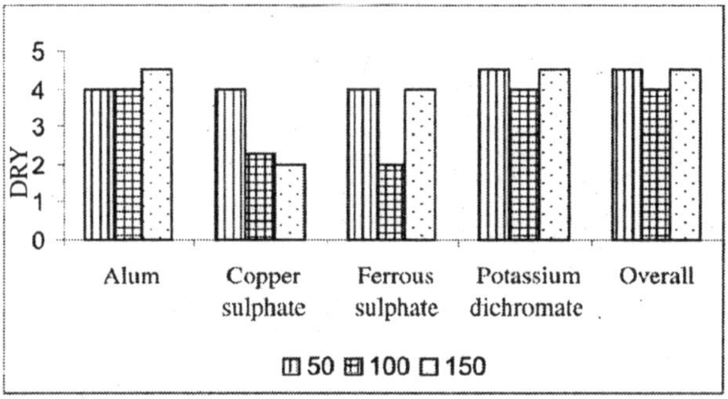

	50	100	150
Alum	5	4.5	4.5
Copper sulphate	5	4.5	4.5
Ferrous sulphate	4.5	4.5	4.5
Potassium dichromate	4.5	5	5
Overall Grading	5	5	5

RED GRAM LEAF
WASHING FASTNESS—CHANGE IN COLOUR

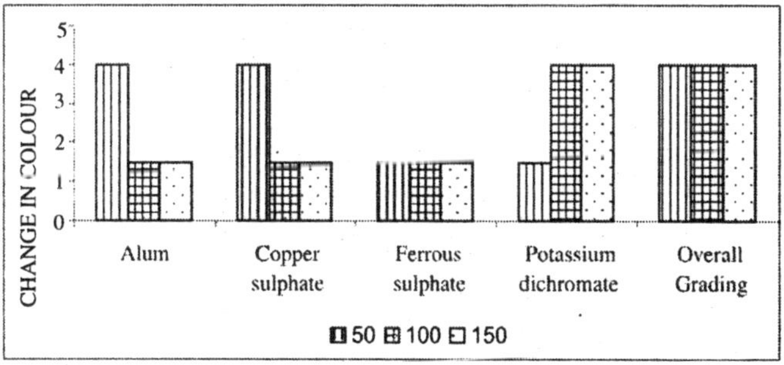

	50	100	150
Alum	3	3.5	2
Copper sulphate	3	2.5	3.5
Ferrous sulphate	4	2	1.5
Potassium dichromate	1.5	3	3
Overall Grading	3	3.5	3

RED GRAM LEAF
RUBBING FASTNESS—WET CONDITION

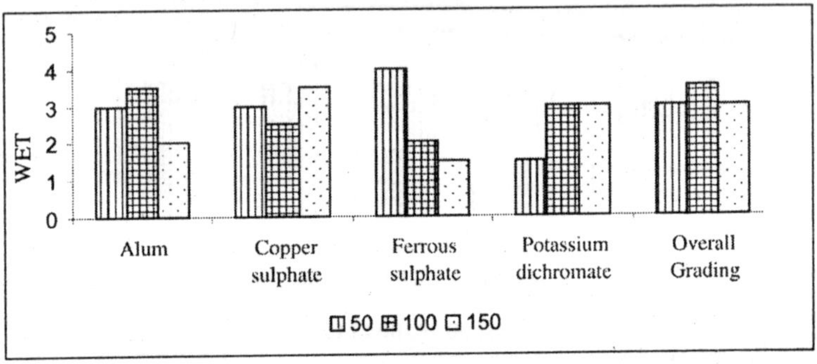

	50	100	150
Alum	4.5	4	4
Copper sulphate	3.5	4	4.5
Ferrous sulphate	4	3.5	4.5
Potassium dichromate	4.5	4	4
Overall Grading	5	4	4

RED GRAM LEAF
WASHING FASTNESS—STAINING ON COTTON

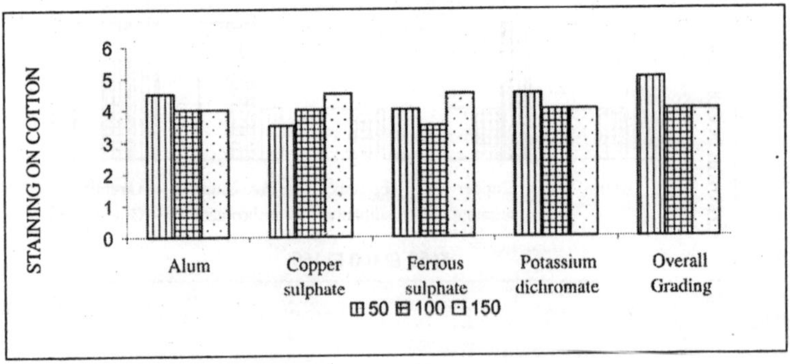

Among the four metallic mordants potassium dichromate has shown higher fastness rating than the other three mordants in respect of change in colour in washing test.

All the other three mordants have also recorded fairly good fastness rating as 4/5 confirming that red gram leaves dye yield fairly good fastness with washing test. The overall rating ranged from 5 to 4/5 with hightest rating at 50 per cent concentration of the dye and at 150 per cent concentration of the dye source. 100 per cent yielded fairly 4/5 fastness rating.

1.6 Wash Fastness Test (Staining on Cotton)

The wash fastness test-staining on adjacent white fabric has yielded fairly good fastness with red gram leaves dye sources.

Among the mordants, ferrous sulphate has shown higher fastness rating of 4/5 than alum, potassium dichromate and copper sulphate.

Overall fastness rating was good at 50 per cent concentration of the dye source and 100 per cent and 150 per cent concentration have very little effect in improving the fastness rating of the dye on cotton fabric.

Further, the overall grading in this test has shown that 50 per cent concentration yielded 4/5 fastness rating followed by 100 per cent and 150 per cent concentration yielding a grade of 4 on a 5 point grey scale rating. Fastness graphs were shown in G2.

On the whole red gram leaves have recorded better fastness in above two tests than Indian mulberry roots and this could be a valuable dye source converting the agro waste into wealth.

Summary of the Findings

The present exercise was undertaken to evaluate the viability of red gram leaves as a vegetable dye source that is widely cultivated or available in Dindigul District of Tamil Nadu. The major dye source selected for the present work is

Red gram plant leaves (Cajanus cajan)

As the vegetable dye matters do not have any direct affinity to cotton fabrics, tanning and mordanting with metallic salts are done. Different concentrations of the dye extract were selected at 50 per cent, 100 per cent and 150 per cent. Different mordants behaved and yielded different colours /hues with the same dye sources.

The cotton goods are dyed in the laboratory as per the dyeing plan. The summary of the finding are given below based on the dyeing trials and fastness testing done.

Performance of Red gram leaves (Cajanus Cajan)

Red gram leaves may be used as a dye source to dye cotton materials to get shades in brownish yellow, yellowish brown, rock brown and reddish khaki. The overall fastness grading varied from 4/5 to 3. 50 per cent and 150 per cent concentration yielded good fastness in dry rubbing condition. On the whole, red gram leaves have recorded better fastness in above two tests and this could be a valuable dye source converting the agro waste into wealth.

Due to the harmful effects of the synthetic dyes, vegetable dye sources find increasing usage and the above discussed findings will help the dye houses and other interested groups to try the dyes on cotton textiles and other materials made out of cellulosic textile materials.

Conclusion

All the four vegetables dyes discussed in the present work demonstrated fairly good to moderate fastness rating on cotton garments. They could be used as commercial dyes if the trials are conducted with some more dye fixing agents that help in fixing the dyes at faster rate.

Further trials are also necessary in reducing the concentration for economising the dye consumption. Trials on dye uptake percentage have to be undertaken among the above dyes to test its dyeing behaviour. Some of the results are quite new through this study which could be used for further application. These vegetable dye sources could be applied on cotton garments with Batik prints.

REFERENCES

Compendium of Inter-Refunel workshop on Natural Dyes, ed. Q.M. Humayen, NHDC, Lucknow.

John and Margaret Canon, *Dye Plants and Dyeing,* The Herbert Press, London, V.K.

Nanda *et al.* Utilisation of Natural Dyes from Plant Wastes. 'Convention proceedings of Natural Dyes', I IT, New Delhi, Dec. 2001.

R. Seerangarajan & S. Jayapal, Fastness properties of Natural Dyes - An assessment. 'Convention proceedings of Natural Dyes', IIT, New Delhi, Dec.2001.

6

BOMBAX MALABARCIUM DYE EXTRACT

S. Amsamani

Eco friendly, environment safe natural products are the need for the day and it is so for textiles too. The biggest damage to the environment is supposed to be done by the dyeing units. It is known to generate many waste streams including water based effluents as well as air emission, solid waste and hazardous waste. Therefore, the key elements for preventing pollution are adoption of natural dyes.

Unlike synthetic dyes, natural dyes are harmless to man and the environment. They are least toxic; contain non-polluting waste effluents and less health hazardous. In nature certain products are left as plant waste and they could be used for dye extraction.

Silk, 'The Queen of Textiles' has been an separable part of Indian culture and tradition. Therefore, if one could dye silk using natural dyes, we could have colourful celebration without polluting our green globe.

Bombax Malabarcium is commonly named as cotton tree. It is also known as Burger, Seemapula, Elayam, Sayar, Simbal, Elavamarum, Mullilary, Illava Burugar. The cotton obtained from its pods are used for stuffing, flower for cooking and trunk for making tea boxes, leer as fuel.

In order to find a better utility valve for the trunk an attempt has be made to optimize a procedure for dye extraction from wet and dry barks of Bombax Malabarcium and to apply the same on mulberry silk.

Experimental Procedure

Selection of Dye

Natural dyes are dyes and pigments that are obtained from plants, animals, vegetable and mineral sources without chemical processing. Natural dyes are being looked as an ecosolution and eoo-friendly dye. Therefore, natural dye was selected for the study.

Selection of Mordants

Mordant are also known to provide a variety of colours with the same dye. Synthetic mordant cause effluent problems and also affect the environment. Therefore, natural mordant are safe to use. Pomegranate rind can also be used as an mordant to get more deeper shade which has good fastness properties. Therefore, eco-friendly mordants, namely myrobalan and pomegranate rind were selected for the study.

Preparation of Fabric

Silk fabric can to degummed by the solution having 0.5 ml genteel in 100 ml of water and stirs gently for about 30 minutes state Rawat *et al.* (2003). Considering the above facts, 25 ml of genteel was mixed in five litres of water. Purchased six metres of silk was soaked for five minutes in cold water, as a means of wetting. Then the water was squeezed out. Later the fabric was soaked in the prepared solution for 30 minutes. The fabric was stirred gently, later it was rinsed twice, thoroughly in cold waters and dried in shade.

Optimization of Dye Extraction

Dry bark of Bombax malabarcium. The wet barks were broken into small pices and grinded into a smooth paste. Dry barks were made into fine powder by grinding the bark in flourmill. Half a kilogram of Bombax malabarcium wet paste and dry powder were taken in four separate vessels and mixed with 1000 ml of water. The solution was brought to boil by placing the vessel on a gas stove. The dye liquor was extracted, at three different temperature 80°C, 100°C, 10°C. The temperature was maintained by increasing and decreasing the flame. The dye extraction time was optimized for 60°C inutes and was kept constant for all the above temperature. After the stipulated time each of the solutions were filtered separately and kept in clean glass bottles.

One ml of each of the dye extract was taken separately in a beaker; it was mixed with 100 ml of water. One gram of silk material was dropped into each of the dye solution and brought to boiling temperature. The dyeing time was kept constant as 30 minutes. Later the dyed samples were rinsed in cold water and dried in shade. The tear strength of each of the samples was noted. The samples were labelled and pasted neatly on a white chart paper and displayed for visual inspection to analyze luster, texture, depth and evenness in dyeing. The results proved that the dyes extracted from 1kg of dry and wet bark of Eucalyptus were good. Therefore, the dyes extracted from one kilogram dry and wet bark of Bombax malabarcium were selected for the study.

Optimization of Dyeing Temperature and Time

Three, six and nine per cent of the extraction from one kilogram Bombax malabarcium dry and wet bark were taken in six separate beaker respectively 100 ml of soft water was added and one gram of silk material was dropped into the solution. Each of the dye bath was maintained at 80°C, 100°C and 120°C respectively. The dyeing time was kept constant as 60 minutes. Later the samples were washed thoroughly and dried in shade. The samples were tested for appearance and strength. The results proved that the samples dyed in 100°C temperature to be good in appearance and strength irrespective to the dye used. Therefore, the dyeing temperature was standardized as 100°C.

Optimization of Mordant Concentration and Dye Concentration

To optimise the dye concentration the silk samples were dyed in 3 per cent, 6 per cent and 9 per cent of dye extracted from Bombax malabarcium in their wet and dry condition, respectively.

Selection of Mordanting Technique

The fastness of a mordant depends on the mordant and mordanting method because different metal dye complexes are formed which may differ in their stability to light and also because the metal may have a positive or negative effect on a photochemical degradation of dye remarks Gupta (1999). Therefore, mordanting was carried out by three methods premordanting, simultaneous mordanting and post rnordanting.

Pre-mordanting

In pre-mordanting method, the degummed material is mordanted for 60 minutes at 100°C in solution containing 2 per cent of pomegranate rind and dried in shade, without any intermediate washing. Dyeing has been done in solution containing 3 per cent of Bambax malabarcium for 60 minutes at 100°C temperature. After dyeing the samples were removed from dye bath and rinsed under running water to remove excess dye particle present on the surface. Similar method was followed for 6 per cent and 9 per cent of dye extract.

Simultaneous Mordanting

In simultaneous method, the solution containing 3 per cent of Bombax malabarcium dye, 2 per cent of pomegranate rind and degummed material were boiled at 100°C for 60 minutes. After dyeing the samples were rinsed under running water to remove excess dye particle present on the surface. Similar method was followed for 6 per cent and 9 per cent of dye extract.

Post-mordanting

The degummed sample was dyed in solution containing 3 per cent of Bombax malabarcium dye at 100°C and then left in mordanting solution containing 2 per cent of pomegranate rind for 60 minutes at 100°C. After dyeing the samples were removed from dye bath and rinsed under running water to remove excess dye particle present on the surface. Similar method was followed for 6 per cent and 9 per cent of dye extract.

After Treatment

As a means of after treatment, the dyed samples were kept in 5 per cent detergent solution for 15 minutes later the material was rinsed three to four times in cold water thoroughly and dried in shade. Following the same process the same were also dyed with 2 per cent of myrobalan as a mordants.

The dyed samples were visually evaluated based on luster, colour, depth of shade, and evenness in dyeing. The results proved that silk dyed in Bombax malabarcium of dry bark in 6 per cent of dye with pomegranate rind in post-mordanting were good and satisfactory.

Actual Dyeing

In actual dyeing 480 grams of dry dye powder of Bombax malabarcium was taken with 14 litres of soft water in a vessel, the solution was brought to boil by placing the vessel on a gas stove. The dye liquor was extracted at 100°C temperature for 60 minutes. The extracted dye liquor was filtered using a muslin cloth. The dyeing time was 60 minutes. Pomegranate rind powder weighing 288 grams was added for post-mordanting. The dyed material was removed from dye bath and rinsed under running water to remove excess dye particles present on the surface.

Both the dyed material were dried in shade and later boiled in soap solution for 15 minutes. After soap-boiling the material was rinsed three to four times in cold water thoroughly and dried in shade.

Wear Study

The dyed fabrics were converted into midi top and scarf, later they were subjected to wear by suitable subjects. This procedure was carried out for 15 days. The samples were washed after every wear by using reetanut.

Evaluation

Evaluation of samples were tested both subjectively and objectively.

The samples were named as SO, SBP_1 and SBP_2 for the original dyed and washed materials.

Subjective Evaluation (Visual Evaluation)

A panel of 25 Post-graduate students specialized in the field of Textiles and clothing were selected as judges for evaluating the sample for general appearance, colour, evenness of dyeing and luster.

Objective Evaluation

Objective evaluation includes laboratory tests.

Results

Visual Evaluation

The visual evaluation reveals that for general apprearance almost all samples were rated to be good by more than 95 per cent of judges. For brilliancy of colour it was rated as bright by more than 93 per cent of judges. As regards to evenness of dyeing it was rated as even by 98 per cent of judges. In case of texture it was rated as soft by more than 93 per cent of judges and for luster it was rated as high by more than 95 per cent of judges.

Thus, from the visual inspection of samples it can be concluded that in general all samples were good in general appearance, bright in colour evenly dyed with soft texture and high luster.

Fabric Weight

The results of fabric weight of original dyed, dyed and washed samples are presented in the Table below:

Table 6.1 : Fabric Weight of Original, Dyed, Dyed and Washed Samples

S.No	*Samples*	*Mean (Ounces per square yard)*	*Std. Deviation*	*Loss or Gain*	*% Loss or Gain*
1.	Bo	16.4	0.52		
2.	SBP_1	19.2	0.63	2.80*	17.1
3.	SBP_2	19.6	0.52	3.20*	19.5

* Fabric weight increased by 17 and 20 per cent after dyeing and washing respectively.

Fabric Stiffness

The results of fabric stiffness of original, dyed, dyed and washed samples are presented in the Table 6.2.

Table 6.2 : Fabric Stiffness of Original, Dyed, Dyed and Washed Samples

S.No	*Direction*	*Samples*	*Mean (cms)*	*Std. Deviation*	*Loss or Gain*	*% Loss or Gain*
		SO	1.9	0.12		
1.	Warp	SBP_1	1.31	0.11	-0.59	-31.1
		SBP_2	1.43	0.09	-0.47	-21.1
		SO	1.9	0.11		
2.	Waft	SBP_1	1.8	0.10	-0.07	-3.7
		SBP_2	1.8	0.11	-0.32	-17.0

* Fabric stiffness reduced in the warp and the weft direction after dyeing and washing.

Breaking Strength

The breaking strength of elongation and load increased by 37 per cent and 13 per cent in the warp and weft directions respectively.

Fabric Thickness

Fabric thickness of the dyed sample increased by 38 per cent and wash sample showed an increase of 42 per cent.

Crease Recovery

The crease recovery of dyed acid washed samples increased by 5 and 10 per cent respectively in warp direction. Whereas the crease recovery of dyed and washed samples decreased by 12 and 11per cent respectively in the weft direction.

Drapability

The results of drapability of original, dyed, dyed and washed samples are presented in the Table 6.3.

Table 6.3 : Drapability of original, dyed, dyed and washed samples

S.No	*Samples*	*Mean (grns)*	*Std. Deviation*	*Loss or Gain*	*% Loss or Gain*
1.	SO	17. 3	3.52		
2.	SBP_1	27.7	0.42	10.40*	40.1
3.	SBP_2	29.8	0.42	12.50*	72.3

* Drapability silk material increased after dyeing and washing.

Tear Strength

Tear strength of silk fabric was decreased after dyeing.

Colour Fastness

All the dyed samples show very good colour fastness to sunlight, crocking, pressing, washing and perspiration.

Wear Study

The dyed, dyed and washed samples did not create any health hazards to the wearer.

Conclusion

From the results of the study it could be concluded Bombax

malabarcium barks could be effectively used for extraction of dye. Apart these dyes could be applied on pure mulberry silk fabrics, which could be successfully used as a apparel. The evaluation of the dyed fabrics also reveal the fact that they have good breaking strength, drapability, fabric stiffness, crease recovery, tear strength and colour fastness properties.

REFERENCES

Gupta, D., 1999, "Fastness properties of Natural dye", *Colourage,* Vol. XLVII, No.3, p. 23.

Rawat, Bhawana, Jahan, Shahnaz, Yadav, Smriti, 2003, "Optimization of dyeing variables for Euphorbia Leaves on Silk," *Indian Silk*, Vol.32, No.9, p. 23.

7

CLONAL FORESTRY PLANTATIONS FOR INDUSTRIAL ESTABLISHMENT

C. Sekar, G. Somasundaram, Anoop G. and *M. Anjugam*

Introduction

The proportion of forest area to national land area in India was about 40 per cent 100 years ago but declined to about 19 per cent in 1993 (PED, JICA, 2002). The Forest Survey of India (FSI) in its assessment presented in the State of Fort Reports (SFR), after monitoring on a biennial cycle using remote sensing data, illustrated that the country's forest cover is 20.55 per cent and the tree cover is 2.48 per cent, together constituting 23.03 per cent of the country's total geographical area (Government of India, 2001). Albeit the National Forest Policy (1988) has stipulated that 33 per cent of the total geographical area should be under forest cover, the assessment revealed that the actual forest cover is much lesser than that of the one enunciated in the forest policy. The per capita forest area in India is also much less *viz.,* 0.08 ha against the global average of about one ha (FAO, 2000). However, it throws in a great deal to the national economy, which is arrived to 1.8 per cent of the GNP. In India the demand for wood increased over years due to faster Industrialization and Urbanization. It has been estimated that the demand of raw materials for the paper and pulp industry will grow from 6.3 lakh tonnes in 1991-92 to 15.5 lakh tones in 2000 and 46.5 lakh tonnes by 2015 (Shanmugam *et al.* 2002). To meet the varied requirements of the burgeoning population, trees are being cut at enormous rates. This indiscriminate felling of forest trees and the usage of forest products, in the recent years, has embarked on to impinge directly upon the economic and environmental health of the nation.

As the forest cover declines more rapidly, the call to safeguard the environment has come from all quarters. Governments have come out with guidelines to counter the colossal devastation of forests. Ultimately, the forest department's prior objective of meeting the needs of the public has been changed to upholding the ecological balance. However, this has condensed the accessibility of raw material to key consumers such as paper and plywood industries and for fuel wood requirements. Even though alternatives are being considered to meet the industrial requirement of pulp wood, the major emphasis still relies on supplementary measures of innovative and scientific advancement to enhance the wood yield over the traditional planting. The prominent supplementary practices are macro and micro propagation of soft wood and hard wood tree clones on a wide scale and subsequent scientific horticultural management.

Most of the Industries are adopting agroforestry as an alternative land use, become a key point of the plantation programmes oriented towards the small farm operation. In this respect, the industrial establishments can come forward to establish the tree plantations in the premises of their industries. In this context, the wood based industries have been forced to choose their options either by having tie up arrangements with farmers to raise the plantations on their own in their industrial premises. The farmer arrangement promoted agro forestry practices in the farm lands and the later one called to raise forestry clonal plantations. Before discussing the dimensions clonal forestry plantations an insight into the pollution load released by the wood based and other industries who are causing damages to the environment are found to be important.

Industrial Pollution in Tamil Nadu

There are several Industries who are seldom use the waste water for different purposes. Instead, they let into various reverine sources and the rivers get polluted thereby causing a threat to aquatic and terrestrial living forms. The details of industries who are releasing pollution lead into the rivers of Tamil Nadu are furnished in Table 7.1.

Table 7.1 : Details of Industries Releasing Pollution Lead into the Rivers of Tamil Nadu

Sl. No.	*Type of Industry*	*Number of Units*	*Quantity of Effluent Discharged in KLD*	*Pollution Lead in Kg / Day*			
				Before Treatment		*After Treatment*	
				BOD	*TDS*	*HOD*	*TDS*
1.	Pharmaceutical	1 (0.25)	345 (0.22)	22	749	3.7	530
2.	Textile processing	357 (88.37)	41590.3 (26.35)	5434.91	184041.6	2214.79	127573.7
3.	Man Made Fibre	1 (0.25)	32500 (20.59)	357.5	65	65	357.5
4.	Chemical	2 (0.5)	390 (0.25)	10 5	0	0.35	0
5.	Electro Planting	1 (0.25	0 (0.00)	0	0	0	0
6.	Pesticides	1 (0.25)	0 (0.00)	0	0	0	0
7.	Tann	14 (3.47)	853.7 (0.54)	583	20145	110.81	5487.2
8.	Thermal Power	1 (0.25)	75000 (47.52)	0	0	75	43800
9.	Sago	23 (5.69)	1537 (0.97)	4007.1	2675	40.14	2675
10.	Milk Chilling	1 (0.25)	10 (0.01)	0	0	0	0
11.	Sugar	1 (0.25)	650 (0.41)	1453	748	82	621
12.	Pulp and Paper	1 (0.25)	4950 (3.41)	1930	11880	891	7920
	Total	404 (100.00)	157826 (100.00)	13398.01	220303.6	3482.79	188964.4

Source: State of Environment Report of Tamil Nadu, 2002

Table 7.1 revealed that 12 types of industries are polluting the riverine systems. The type of industries who are considered as giant in polluting the riverine environment are Textiles processing followed by Sago industries and Tanneries, accounted for 88 per cent, 5.69 per cent and 3.47 per cent respectively to the total number of industries polluting the riverine environment.

Regarding quantity of effluent discharged in kilolitres per day (KLD), Thermal Power ranked first who shared 47.52 per cent of total discharge into the riverine environment followed by Textile processing and Man-made fibre shared 26.35 per cent and 20.59 per cent respectively. The total discharge of pollution load was accounted to be 1.58 lakh kilolitres per day by 404 industries spread over in the length and breadths of different rivers of Tamil Nadu.

Table 7.1 also revealed that the pollution load had been lowered to a great extent on treatment of polluted waste water released by the Industries. The Biological Oxygen Demand (BOD) and Total Dissolved Solids (TDS) have been lowered and thus, it can be made use of for different activities, viz. agriculture, forestry, fishery purposes or in combination of these.

Here, an emphasis was given to highlight the feasibility of establishment of clonal forestry plantation to meet the raw material requirement for pulp and paper industries present in Tamil Nadu.

Clonal Forestry—An Overview

Clonal forestry, proposed in 1906 as a general system of tree breeding programme, has emerged as a strong alternative to the conventional seed orchard breeding system and is considered to have vital importance in tree improvement. Clonal forestry is a method to mass produce desired trees and to get the desired benefits by using outstanding and known good genotypes. It does not create anything new or directly produce improvements in advanced generations. But it enables taking advantage of differing genes through maturation and using the knowledge gained from clonal testing, which indicates the value of the genotype when it is used as a clone. Clonal propagation was widely made use of for the preservation of genotypes in clonal banks and for clonal seed orchard establishment. Techniques were developed for large scale propagation of many forest tree species from cuttings at costs on par or even less than that of seedlings. Recently, there is a growing interest in using the techniques of clonal forestry for exploiting the considerable amount of genetic variability already existing in the natural populations of forest tree species that are amenable to vegetative propagation (Nanda, 1970) and because of the tremendous yield improvement by adopting the clonal forestry approaches (Pal, 1991).

Clonal forestry has proved its superiority to seedling propagated trees as the clones are selected based on their performance as 'reliable' clones. Once identified, high-yielding, high-quality clones can be quickly deployed in large numbers per clone for practical forestry purpose; thereby avoiding the disadvantages associated with sexual recombination in seed orchards. Further, it offers the possibilities of using specific adaptations or trait combinations, giving the breeders and silviculturists greater flexibility.

In Tamil Nadu, the list of plus trees, otherwise called as Candidate Plus Trees (CPT) selected for clonal propagation with respect to *Eucalyptus*, *Casuarina* and *Teak* propagation are furnished in Table 7.2. However, apart from a few genera like *Poplar, Salix, Cryptomeria, Eucalyptus* and *Casuarina,* clonal propagation has not been used extensively in operational forest planting programmes because of poor awareness among the planters and inadequate research and development activities.

Table 7.2 : Candidate Plus Trees in Tamil Nadu

Sl. No.	*Species*	*No of plus trees*
1.	*Casuarina equisetifolia* (male)	36
2.	*Casuarina equisetifolia* (female)	115
3.	*Eucalyptus globulus*	40
4.	*Tectona grandis*	24

Source : Tiwari, 1994.

In this context, effort has been taken to appraise the economic incentives associated with the production of clonal plants, micro propagules and practicing clonal forestry.

Data Collection and Analysis

The costs of requisite raw materials, processing of clonal cuttings, media development, rooting and hardening were acquired from various sources, viz., Forest College and Research Institute, Mettupalayam as well as Institute of Forest Genetics and Tree Breeding, Coimbatore. The organization of total annual costs of generating clonal plants and plantlets by tissue culture is represented by the following expression used by Pachauri and Dhawan (1989), Sekhar *et al.* (2000).

$$T_c = (K_c + K_b)(i+d)+E+S+L_k$$

where,

T_c = Total annual cost of production
K_c = Total cost of equipment
K_b = Total cost of building and infrastructure
i = Annual rate of interest applicable to investments
d = Annual rate of depreciation
E = Annual cost of power consumed
S = Annual cost of consumables and supplies
L_k = Annual cost of labour in k^{th} process activity
k = an integer denoting the number of activities covered in the process such as collection and preparation of explants, preparation of nutrient media, rooting media, etc.

The above method was adopted for estimating the total cost of production per unit quantity. L refers to the labour wages for initiation of cultures, collection of clones (L_1), shoot multiplication (L_2), rooting (L_3), hardening (L_4) and transplantation (L_5). Regarding the production and productivity of clonal and tissue cultured plantations of *Eucalyptus tereticornis*, the yield of four year old tissue cultured plantations of Tamil Nadu Forest Plantation Corporation was extrapolated (Gupta, 1981) to the harvesting age of five years for comparison with the existing five year old seedling plantations. The unknown values were derived by extrapolation using the Binomial Expansion Method (BEM) with the assumptions that there are no sudden jumps in the series from one period to another and the rate of change of figures from one period to another is uniform.

Economics of Clonal Forestry

Economic appraisal taken up in this chapter involves Investment Analysis, Cost-Volume-Profit (CVP) Analysis, Break Even Analysis and Capital Efficiency Analysis.

Investment Analysis

Investment analysis drawn in various estimates as mentioned below :

Fixed Cost

Salary to permanent staff, if any, depreciation on buildings, mist chamber, shade net house, machinery and equipments, imputed rental value of land, interest on fixed capital and property tax were incorporated in the fixed cost. Same buildings, same mist chamber and shade houses were used for producing clonal plants for a number of years. Hence, the fixed cost was apportioned to different plants produced considering the number of days required for processing for each clonal plant. The apportioned fixed cost was arrived at Rs. 16,775 for a production period of six months. These are presented in Table 7.3.

Variable Cost

The components of the variable cost were worked out based on the operations followed in the production of clonal propagules. Various activities in the production activity of clonal plants are collection of superior clones, cutting, trimming, processing with growth hormone, media development, fertilizer and plant protection chemical spray, watering, electricity and other miscellaneous costs. The variable costs were worked out, operation-wise were summed up to arrive at the total variable cost, viz., Rs. 15,215 for producing 10,000 clonal plants with 70 per cent success rate.

Table 7.3 : Cost of Production of Clonal Plants of *Eucalyptus sp.*

Sl. No.	*Particulars*	*Actual cost incurred (Rs.)*	*Life span of asset (years)*	*Cost per 10,000 clones (Rs.)*
A.	**Fixed Cost**			
1.	Investment on buildings	80,000	20	2,000
2.	Mist chamber with automatic control unit (60 sq.m)	1,50,000	20	3,750
3.	Shade net house with micro sprinkler and fogger	80,000	10	4,000
4.	Root trainer stand	30,000	10	1,500
5.	Root trainer pot @ Rs. 7 per pot	70,000	10	3,500
6.	Farm equipments	3,000	3	500
	Total Fixed cost	**4,13,000**		**15,250**
7.	Interest on fixed investments @ 10 per cent	41,300		1,525
	Fixed cost+ Interest on fixed investments	**4,54,300**		**16,775**
B.	**Variable Cost**			
1.	Electricity			1,440
2.	Annual maintenance of fixed structures @ 1 per cent of total fixed cost	4,450		2,225
3.	Collection of clones (6 man days)			480
4.	Trimming and processing with IAA/IBA			1,600
5.	Rooting medium (Vermiculite)			8,120
6.	Macro and micro nutrients + plant protection chemicals			200
7.	Nutrient and pesticide spray @ 1 man day per month			480
8.	Weeding to clonal plants (3 man days)			240
9.	Transport and miscellaneous charges			430
	Total variable cost			**15,215**
	Total cost (A+B)			**31,990**
C.	**Cost of production per clonal plant assuming 75 per cent success rate**			**4.27**
D.	**Sale price of clonal plants**			**7.00**
E.	**Gross income**			**49,000**

Total Cost

The total cost is the sum of total fixed cost and total variable cost incurred in the production activity of clonal plants. Here, the total cost was computed to Rs. 31,990 in which the total variable cost alone was Rs. 15,215 which accounted for 47.56 per cent to the total cost.

Gross Return

The gross return was computed by multiplying the quantity of successful clonal output by the selling price per clonal plant. Gross return was worked out using the average price that are prevailing with the institutions like Forest College and Research Institute, Mettupalayam, Institute of Forest Gentics and Tree Breeding, Coimbatore and the Badrachalam Paper Boards (Private) Limited. The prices prevailing in these institutions respectively were Rs. 4, Rs. 10 per clonal plant (Eucalyptus). The average price per plant was worked out to be Rs. 7 and the same was used to compute the gross return. Here, the gross return was worked out to be Rs. 49,000. This could be reaped in a period of six months.

Net Return

Net return was arrived at by subtracting the total cost from the gross return. In the production activity of clonal plants, net return was arrived at Rs. 17,010 in a production period of six months.

Since in both Euacalyptus and Casuarina softwood cuttings are drawn from elite clones, fixed inputs, operational expenses and pricing are roughly the same and hence the cost of production arrived for Euacalyptus is treated as equal for Casuarina. Hence, the cost of production per clonal plant of Euacalyptus, viz., Rs. 4.57 holds good for Casuarina also (Table 7.3).

However, comparison was made to highlight the cost of production of clonal *Euculyptus sp.* plants *via-à-vis* tissue cultured plantlets of *Eucalyptus sp.* and the costs are furnished in Table 7.4.

Table 7.4 revealed that the cost of production per tissue cultured plantlet of *Eucalyptus sp.* is Rs. 4.65; the total cost of production of one million plantlets being Rs. 46.46 lakhs. Among this, fixed costs accounted for 30.80 per cent of the total cost, while variable costs involved were arrived at 69.30 per cost to the total cost. The

Table 7.4 : Cost of Production of One Million Tissue Cultured Plantlets of *Eucalyptus sp.*

Sl. No.	*Particulars*	*Amount (in Rs)*	*Percentage to total cost*
A.	**Fixed Cost**		
1.	Building and infrastructure establishment	2,00,000	2.1
2.	Equipments	10,25,000	24.6
3.	Interest on investment	1,12,500	2.7
4.	Depreciation	58,188	1.3
	Total fixed costs	**13,95,688**	**30.8**
B.	**Variable Cost**		
1.	Power	10,25,000	22.0
2.	Consumable and supplies	4,70,000	9.9
3.	Labour wages for initiation of cultures	3,41,000	6.4
4.	Shoot multiplication	5,90,000	13.1
5.	Rooting	3,00,000	5.9
6.	Hardening	3,40,000	8.3
7.	Transplanting	1,85,000	3.9
	Total variable costs	**32,51,000**	**69.3**
	Total cost (A+B)	**46,46,688**	**100.0**
	Cost of production per plantlet assuming 90 per cent success rate	**4.65**	

cost of production of tissue cultured plantlet is almost equal to the cost of production of clonal plants. The production cost per seedling generated from seeds was Rs.0.80 (Table 7.5) whereas, clonal and tissue cultured plants were incurring a cost six times higher. However, amidst the higher costs, the clonal and tissue cultured plants have yielded considerable returns. Hence, what is required is reduction of failure rate of clonal output by taking stringent efforts during processing of clones. Infection also needs to be prevented at all stages.

Cost-Volume-Profit (CVP) Analysis

The Cost-Volume-Profit (CVP) analysis measures the effect of changes in volume, cost, price and product-mix on profits. Costs vary due to changes in choice of clonal plant, scale of operation,

Table 7.5 : Cost of Production of Plantlets of *Eucalyptus sp.* through Seed Origin

Particulars	*Amount (in Rs.)*	*Percentage to total cost*
Establishment cost	2,92,000	36.6
Maintenance cost	4,14,000	51.9
Rental value of land	19,000	2.4
Interest on investment	72,500	9.1
Total cost	7,97,500	100.0
Cost of production per seedling	0.80	

technology, efficiency of work-force and management efficiency. Though many wide ranging factors influence costs and profits in any production activity, the largest single variable affecting them in the short-run production is the volume of clonal output. Hence, CVP relationship acquires a vital significance for the farm manager facing a wide spectrum of short-run decisions like: what are the most successful and profitable species with respect to consumers' preference; how does a reduction in selling prices affect profits; what will be the break-even-point if volume and costs change; how an increase in wages and/or other operating expenses will affect sales of clonal plants. Answers to all such questions will have to be formulated in a benefit-cost framework and CVP analysis will offer the technique of doing it. CVP analysis is in fact one of the decision-models which plantation managers employ to choose among alternative courses of action. The basic (simplified) CVP model may be outlined as follows:

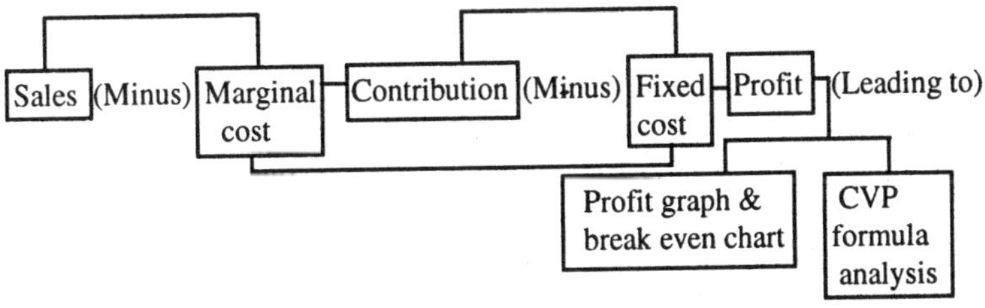

Fig. 7.1. Division of costs into their variable and fixed cost elements

CVP concept comprehends that profits are a function of the interplay of costs and prices and each one of them is relevant to profit planning. In fact, variance between actual and budgeted profit is due to one or more of the factors, viz. selling price, volume of sales, variable costs, and fixed costs. It is obvious that selling prices largely depend upon external forces. Costs, of course, are more controllable. However, they pose a problem of measurement. This is more so when the farm produces two or more tree species. Nevertheless a knowledge of fixed and variable costs are essential, if costs are to be controlled and reduced.

CVP analysis was exclusively carried out to find the cost-volume-profit relationship in the clonal plantations. Here the CVP analysis was carried out in respect of bulk plantations raised through seed origin was compared with the clonal and tissue cultured plantations of Eucalyptus (Table 7.6).

In respect of Eucalyptus plantations, the bulk plantations are compared with the clonal and tissue cultured plantations. Among the three plantations tissue cultured plantations had generated 91 tonnes per ha which was accounted to be 1.75 times higher yield than the bulk. Whereas, the increase was only 1.65 times in respect of clonal plantations which is comparatively lesser than the tissue cultured plantations. Since these plantations absorb unique maintenance and other costs, the major difference in cost was the

Table 7.6 : CVP Analysis of Eucalyptus Plantations in Tamil Nadu

Sl.No.	*Particular*	*Eucalyptus*		
		Bulk	*Clonal*	*Tissue culture*
1.	Spacing (metres)	2 × 2	2 × 2	2 × 2
2.	Age of harvesting (years)	5	5	5
3.	Yield (tonnes per hectare)	52	102 (1.65)	109 (1.75)
4.	Cost of seedlings (Rs.)	5,000	17,500 (3.5)	17,500 (3.5)
5.	Sale price per tonne of pulp wood (Rs.)	1,200	1,250	1,250
6.	Gross income (Rs./ha)	62,400	10,7500	1,13,750
7.	Net income Rs. / ha)	57,400	90,000 (1.57)	96,250 (1.68)

(Figures in parentheses denote the ratio of clonal to bulk values.)

seedlings. Hence the seedling cost alone was taken into account for comparison. The cost was 3.5 times higher than that of bulk seedlings. The net profit was 1.57 and 1.68 times higher than the bulk in respect of clonal and tissue cultured plantations. These called for increased awareness and institutional intervention to augment the wastelands in the industrial premises with clonal and tissue cultured plantations so as to increase the tree cover and at the same to reduce the environmental pollution.

Break Even Analysis

Break-even point = (BEP) is used to determine the minimum level of production that was required to get neither profit nor loss. Here, it was used to determine the quantity of clonal plains that must be produced and sold by each invest to cover the fixed cost. Break-even quantity was arrived at using the following formula.

$$\text{Break-Even Point (BEP)} = \frac{\text{Total fixed cost per production period}}{\text{(Price per clonal plant) - (Variable cost per clonal plant)}}$$

Break-even quantity for production of clonal plant of *Eucalyptus sp.* was arrived at using the average market price of Rs. 7 per plantlet and it was worked out to be 3,474, a quantity that would suffice to reach a point of no gain-no loss. In this case, the production target during a production period (with 70 per cent success rate) was 7000 plants. Hence, there was a production margin of 3526 plants. In respect of tissue cultured plantlet the break even quantity was arrived at 3.6 lakhs and production margin was 6.38 lakh plantlets.

Pay Back Period (PBP)

In simple terms, it means the total period within which the total amount invested will be recovered through the net cash flow. A sum of Rs. 4.54 lakhs was incurred as fixed cost for producing 10,000 clonal plants. The annual fixed cost and the total variable cost together accounted for Rs. 31,990. All these costs could be recovered between fourth and fifth year of production activity and pay back period was 4 years and six month. So establishment of the clonal nursery as an enterprise is found to be highly remunerative.

Discounted Measures of Economic Appraisal

Net Present Value (NPV) and Benefits Cost Ratio (BCR) are reliable estimates of profitability of an enterprise (Reddy and Korwar, 1985, Hoekstra, 1990). Hence these techniques made use of to examine the feasibility of clonal forestry endeavour and highlight the economic incentives associated with it.

Net Present Value (NPV): It is assessed using the formula:

$$NAV = \sum_{i=0}^{n} \frac{Bt - Ct}{(1+r)^t}$$

where, Bt is the benefit obtained in the year 't'

Ct is the cost incurred in the year't'

$i = 1, 2, 3, \ldots n,$

When *NPV* is greater than zero, the plantation was considered to be worthy. Table 7 revealed that the *NPV* was Rs. 28,412 for clonal plantation of *Eucalyptus sp*. These are almost five times greater than that realized in bulk plantations. Hence, establishment of clonal plantations are advantageous.

Benefit Cost Ratio (BCR): It is the ratio between the sum of discounted benefits and discounted costs, expressed as:

$$BCR = \sum_{i=1}^{n} \frac{Bt}{(1+r)^t} \Big/ \sum_{i=0}^{n} \frac{Ct}{(1+r)^t}$$

Benefit Cost Ratio (BCR) was very poor in bulk plantations whereas, clonal plantations have yielded a benefit cost ratio was 1.60 for *Eucalyptus sp*. It revealed that for a rupee of investment, a return of 1.60 could be generated (Table 7.7).

Table 7.7 : Discounted Measures of Economic Appraisal of Eucalyptus Plantations

Sl.No	*Particulars*	*Eucalyptus*	
		Bulk	*Clonal*
1.	Total cost (Rs. per ha)	16,392	30,144 (1.83)
2.	Yield (tonnes per ha)	52	102 (1.96)
3.	Cost of Production (Rs. per tonne)	315	315 (1.00)
4.	Net present value (Rs.) (NPV)	4,908	28,412 (5.78)
5.	Benefit - Cost ratio (BCR)	1.14	1.60

(Figures in parentheses denote the ratio of clonal to bulk values.)

Capital Efficiency Analysis

Capital efficiency measures were worked out to find out the efficiency of capital use in production of clonal plains. Various measure employed here are return on investment, Capital output ratio and Output-input ratio.

$$\textit{Return on Investment (ROI)} = \frac{\text{Net return in a production period}}{\text{Capital investment}}$$

In the production of clonal plants, the production period was 4 to 6 months. On the higher side it was treated as 6 months. During this period the net return obtained from clonal as well as tissue cultured plantlets were Rs. 16,454. Here the return on investment was arrived at Rs 1.02 in respect of clonal plants, whereas, it was Rs. 1.35 in respect of tissue cultured plantlets. The return on investment in clonal plants was poor mainly due to smaller success rate. If the success rate can be increased, a big return on investment is possibly from clonal plants.

$$\text{Capital Output Ratio} = \frac{\text{Capital investment}}{\text{Output}}$$

Capital output ratio was arrived at 0.19, for tissue cultured plantlets of *Eucalyptus sp*. Hence it is better to target the production of clonal at a higher capacity to reduce the per plant investment.

$$\text{Output-Input ratio} = \frac{\text{Output}}{\text{Input}}$$

Output refers to total return or gross return, whereas input refers to the total cost of producing the clonal plants, targeted and capital investment referred to fixed cost. The output-input ratio was arrived at 1.36 with respect to clonal plants and tissue cultured plantlets, respectively. It revealed that for a rupee of investment, the clonal plants could fetch higher return of Rs. 1.53 which was 11 per cent higher than that of the return from tissue cultured plantlet of *Eucalyptus sp*. It was mainly due to the involvement of lessor variable cost for clonal plant production when compared to tissue cultured. To sum up, the capital efficiency measures reported that the return was on the higher side for clonal plant production.

Conclusion and Policy Implications

The chapter on economics of clonal forestry has highlighted the details of costs and returns that are involved in establishing clonal plant and tissue cultured plantlets. The same were compared with the cost of production of seedlings raised from seed origin. It revealed that the cost of production per plant was six times higher than the one involved from seed origin. The yields of clonal, tissue cultured plantations were also showing around 1.75 times higher than that of the bulk plantations. The parameters like Net Press Value and Benefit Cost Ratio had indicated that the clonal plantations are economically viable and could fetch higher returns by expanding the production target of clonal plants in a stipulated production period.

Thus, it called for allocation of higher area for establishment of tree plantations coupled with other agricultural activities so as to effectively use the treated water.

The industrial establishments should allocate one third area for greenery purposes particularly for afforestation activities while advancing the licence for functioning of the industries.

Re-use of water need to be strictly enforced in all industrial establishments so as to protect the aquatic and terrestrial environment from further degradation due to pollution hazards.

REFERENCES

Dewari, D.N., 1992. *Monograph on Neem,* International Book Distributors, Dehradun, p.279.

FAO, 2000. An Abstract of Forestry, *The Report on In-depth Study of Forest Situation in India*, pp. 1-2.

Government of India, 1991: *The State of Forest Report,* Ministry of Environment and Forests, New Delhi, p.19.

Government of India 2001, *The State of Forest Report,* Ministry of Environment and Forests, New Delhi, p. 1.

Gupta, S.P., 1981. *Interpolation and Extrapolation - Statistical Methods,* Sultan Chand and Sons, New Delhi, pp. 15.1-15.31.

Hoekstra, D.A., 1990. *Economics of Agroforestry; Agroforestry Classification and Management,* John Willey Publications, New York, pp. 425-427.

Jayaraj, R.S.C., 1997. Selection of Candidate Plus trees of *Eucalyptus sp.* for a Large Scale Clonal Forestry Programme. in eds. Vivekanandan, Gurumurthi, K. and Jayaraj, R.S.C., *Colonal Multiplication of Eucalyptus,* pp.34-47.

Nanda, K.K., 1970. Investigations *on the* Use of Auxins in Vegetative Reproduction of Forest *Plants, Final Report PL 480 Research Project,* A7 - FS 63.

Pachauri, R.K. and Dhawan, V., 1989. Cost Analysis of Micro-propagated plants, *Application of Biotechnology in Forestry and Horticulture,* Plenum Press: London, pp. 275-282.

Pal, M., 1991. in eds: P. Khullar, R.C. Thapliyal, B.S. Beniwal, R.K. Vakshasys and Ashok Sharma, *Forest Seeds*, ICFRE, Dehradun, pp.343-370.

Planning And Evaluation Department, *Japan International Cooperation Agency*, Country Profile on Environment India, 2002.

Reddy, Y.V.R. and G.R. Korwar, 1985. A Comparative Economic *Evaluation* of

Agroforestry, Silviagriculture and Silvipastoral Systems in Dry Lands, *Agricultural Situation in India,* November, pp.705-708.

Sekhar, C., R.C. Panda and R.S. Vinaya Rai, 2000. Comparative Evaluation of Tissue Cultured Plantation in Tamil Nadu, *Journal of Tropical Forest Products,* 6 (2), pp. 127-133.

Shanmugam, T.R., Palanisamy, N.V. and Mahesh, N. "Measurement of Economic Efficiency in Agroforestry Production System", *Agricultura Tropica EE Subtropica,* Vol. 35, 2002, pp. 25-32.

State of Environment Report of Tamil Nadu, *Government of Tamil Nadu,* 2002.

Tewari, D.N., 1994. *Biodiversity and Forest Genetic Resources,* International Book Distributors, Dehradun, p.329.

8

GREEN PRODUCTIVITY AND SUSTAINABLE INDUSTRIAL DEVELOPMENT

S. Subbiah and *I. Sundar*

Introduction

Green productivity concerns the efficient use of natural resources. The increased efficiency results in the conservation of natural resource and it leads to its sustainable use. Green productivity is essential to protect ecological balance. Usually pollution disrupts the various ecological processes that provide us with clean air and water. It contaminates the food chain that provides human beings with food. Hence green productivity with its objectives of preventing and reducing pollution will protect the ecological processes that are very essential in maintaining the ecological balance of environment.

It is significant to note that plants and animals are valuable in terms of bio-diversity constituents and they are essential for long-term survival of human beings as they confined to help maintain the ecological balance of life support system. They are also the basis of human food and other products. The genetic composition of these plants and animals are genetic resources for the improvement of human food crops, source of medicine and other useful products. In this context, green productivity aims at more efficient use of the environment and resources and the reduction of pollution, which contributes, to the survival of these species.

Green Productivity and Business

Green productivity is a strategy that leads to gain in profitability through improvements in productivity and environmental performance and also for overall socio-economic development. It is the application of appropriate techniques, technologies and management systems, to produce environmentally compatible goods and services. Whereas excessive use of resources are generation of pollution, it induces low productivity as well as poor environmental performance. To improve the situation, green productivity pursues a strategy based on technical and managerial intervention. Thus it is a process of continuous improvements.

Green productivity lays emphasis on resource use efficiency and it concentrates on natural capital. In this context, there is a need to prevent pollution or waste, at its sources as well as reduce the level of resource inputs by the process of rationalization and optimization. Further, the possibilities of reuse recovery and recycling are to be analysed to reduce the waste generation. Efforts should be to substitute toxic or hazardous substances with a view to reduce the lifecycle impact of the product. In this stage a proper examination of anti-environmental impact of the product is required. The next step is to treat waste in residual forms by suitable end of pipe options to meet the regulatory requirements both from the perspectives of the work place and the receiving environment. To ensure a continuous improvement in productivity as well as in the level of environmental protection, a systematic approach and methodology need to be developed for green productivity.

Green Productivity Guiding Principles

Green productivity makes environmental protection while making business profitable. This is a necessary condition to achieve sustainable development. It recognizes the environment and development are the two sides of the same coin. Green productivity makes at ensuring quality by promoting the use of newer and safe materials, increasing, processing and production efficiency and improving working conditions. Thus it provides the consumer with more performance and value with use of less resource, including energy and the creation of less waste.

Accountability of production of goods and service is an essential aspect of green productivity. It emphasizes the need for business to be accountable to various stakeholders. Typically in law, they are accountable to regulators. Further, other stakeholders like the suppliers, customers, consumers and the public at large are increasingly seeking accountability for the actions and impacts of business activity. Polluter pays is another guiding principle of green productivity. It brings the element of responsibility by placing the cost of environmental clean up on the polluter. The rationale is penalization of damages created. It forms the basis of penalties and pollution taxation system. It also makes the polluter responsible for incurring the cost of clean ups. This may be in the form of development of treatment system and pollution prevention technology.

The third guiding principle is precautionary approach. It is typically applicable in situations when the impact of an event is long-term and difficult to reverse. It could be noted that pollution prevention, cleaner production and source reduction are all based on the precautionary approach to environmental protection. Green productivity aims at cost effectiveness. Here profitability is the basis of any business and it advocates the need to generate profits. However it could be achieved through savings in raw materials by practising resource use efficiency, improved productivity, quality sales and so on. Competitive advantage is one of the productivity aspects. It is essential for business to establish and maintain a market position. It emphases the importance of competitiveness in pricing, quality and in the case of green productivity, it is eco-friendliness.

Strategies of Green Productivity

Green productivity aims at preventing waste and emissions at the point of generation. The non-preventable waste and emissions need to be treated to safeguard the environment from over toxic impacts. Waste prevention is one of the aspects of green productivity. Waste prevention techniques can be applied to any manufacturing process, agricultural and service sectors. Good housekeeping is a constituent of waste prevention. It is a process of keeping the work place clean and eliminating leakages and spillages. It also includes the operational practices involving especially manual practices and making them more resource efficient. The activities involved here are preparation of receipts in time, quantity to avoid surplus, efficient handling of materials, optimum storage procedure to avoid losses and material degradation during storage. A good housekeeping may expect to reduce work generation up to 25 per cent.

Inventory management is one of the constituents of waste reduction. It needs a proper control over raw materials, intermediate products, financial product and the associated work streams as an important work reduction technique. In many cases waste is just out of date, off specification, contaminated or unnecessary raw material, spill residues or damaged final products. The cost of disposing of these materials not only includes the actual disposal costs, but also the cost of raw materials and products, which are lost. Inventory control is an important aspect of waste minimization. Some firms can help reduce this waste generation by tightening up and expanding current inventory control programme. This method has its impact on improper inventory control excess, outdate and no-longer used raw material. In this context, a proper material control procedure is required to ensure that raw materials will reach the production process without loss through spills, leaks or contamination. It will also ensure that the material is efficiently handled and used in the production processes and does not become waste. Material loss can be greatly reduced through improved process operation, increased maintenance and additional employee training. Many sources of material loss such as leaks and spills can be easily identified and corrected. Improving the efficiency of a production can significantly reduce waste generation and this approach can help reduce waste at the source of generation. Thus it leads to decrease work management liability and costs.

In order to reduce waste, improvements in the way of production process, operation and maintenance are required. Improvements in operation and maintenance usually are relatively simple and cost effective. In this context, a proper operational procedure is needed. Improved operational procedures optimize the use of raw materials in the process of production. More efficient operation of production process can be achieved through changes, in operational procedure. A proper maintenance programme is needed to reduce about one-fourth of the excess waste.

Change in use of materials in the production process is required. The manufacturing sector uses the most cost benefit inputs raw materials in the production process without considering environmental aspects. Thus, highly toxic and hazardous chemicals came into use. Hazardous materials used either in a product formulation or in a production process can be replaced by a less hazardous or non-hazardous material. Reformulating a product to contain hazardous materials should reduce the amount of hazardous waste generated during product formulation and its end use. Using a less hazardous material in a production process will generally reduce the amount of hazardous waste produced.

Installing more efficient process equipment may reduce waste generation. New equipment can use process materials more efficiently, producing less waste. Higher efficiency systems may also reduce the number of rejected products, thereby reducing the amount of materials, which have to be reworked. In this connection, modifying existing process equipment can be a very cost effective method for reducing waste generation. In many cases, modifications can just be relatively simple and inexpensive changes in the way the materials are handled within the process to ensure that they are not wasted. Further process modifications and improved operational procedures can be used together to reduce waste.

Waste reduction can be done through volume reduction. It includes techniques to separate hazardous wastes and recoverable wastes from the total waste stream. These techniques are usually used to increase recoverability, and to reduce the volume and thus the disposal cost. Source segregation is one of the aspects of volume reduction. It could be noted that segregation of waste is a simple and economical technique for waste reduction. Segregating wastes at the source of generation and handling the hazardous and non-hazardous waste separately can reduce waste volume management cost. The uncontaminated or undiluted wastes may be reusable in the production process or may be sent off site for recovery. The segregation technique is applicable to a wide variety of waste streams and industries and it usually involves simple change in operational procedures. Some concentration techniques are commonly used to regulate wastewater treatment sludges and to reduce the volume by as much as 90 per cent.

The application of waste recovery techniques can help eliminate waste disposal costs, reduce raw materials costs and possibly provide income from a stable work. Waste can be most efficiently recovered at the point of generation, because the possibility of contamination with other waste material is considerably reduced. Other waste streams can be reused directly in the original production process as raw materials. Wastes may be recovered at an off site facility when the equipment is not available to recover on the site, not enough waste is generated to make an inplant systems cost effective and the recovered material cannot be reused in the production process.

Energy Conservation Techniques

Combustion of Fossil fuels in primary heat sources such as boilers or heaters provides a major source of heat input to industrial processes. Taking care to prevent its loss during transport from the combustion site to the specific processes where it is used can conserve thermal energy. Waste generation is inevitable in industrial activities. The waste that cannot be eliminated, reduced, recycled or reused must be treated and disposed of within all applicable environmental regulations. In this context, environmentally, compatible products minimize the adverse effects on the environment resulting from the manufacture, use and disposal. The environmental impact of a product is to a large extent determined during its design phase. Design changes made to prevent pollution should be implemented in such a manner that the quality of function of the product is not affected adversely.

Design for the environment can be achieved by people directly involved within the framework of company policy and with support from company management. Product conservation is one of the mathods to reduce wastes. Product conservation refers to the way in which an end product is used. It is to be observed that better maintenance of process equipment and components by industry can reduce the frequency of equipment component replacement, which in turn brings down the waste generated by the used component. Further product design changes involve manufacturing a product with a lower composition of hazardous substances or less toxic materials being formed or changing the composition, so that no hazardous substances are involved.

Benefits of Green Productivity

Green productivity benefits business by lowering its operational and environmental compliance costs and by preventing the generation of waste through efficient resource utilization. This can also reduce long-term liability and cleanup costs. It also reduces disposal costs in consequence of waste reduction. The shift towards integrating environment and productivity enable by green productivity has much greater implications for business in country like India, which is due to their technological and resource constraints. Green productivity aims at shaping business to answer the social demands for quality of life through supply of goods and services rendered in a sustainable manner. By improving productivity and environmental performance for overall socio-economic development, green productivity enhances the quality of life.

Conclusion and Policy Suggestions

It could be seen clearly from the above discussion that green productivity signifies a new paradigm of socio-economic development aimed at pursuing economic and productivity growth while protecting the environment. Green productivity is applicable not only to the manufacturing sector, but also to the agriculture and service sector. It also addresses the interaction between economic activity and community development. The successful implementation of green productivity programme depends on the following policy measures :

1. A proper formulation green productivity planning machinery and team.
2. Data on environmental situation, physical infrastructure, socio-economic status of the communities should be collected for further analysis.
3. The planners to execute green productivity plan, a proper identification of problem and their causes are required. In this context problems concerning waste supply, waste generated and agricultural activities should be identified. Further problems to be identified depend mainly on the community standard of living, cultural background and to certain extend physical condition of the community.
4. Generation of green productivity option is another aspect like as options in waste management. Alternatives in waste treatment, water supply options and so on.
5. A proper arrangement of technical feasibility. The technical evaluation determines whether the proposed option is technically workable under the given condition.
6. Assessment of economic viability of green productivity is required, since economic viability is often the key parameter for promoting and discussing implementation of green productivity options.
7. The green productivity options need to be analysed with respect to their impacts on the environment.
8. The implementation of green productivity programme requires finance arrangement, technical preparations and establishing linkage with other organizations.
9. A proper implementation of selected options. The task comprises preparation of drawings, ordering and procurement of equipment, transportation of the equipment to the site, installation and commissioning.
10. The green productivity programme implementing officials need a proper training.

11. Monitoring and evaluation of green productivity programme are imperative to test the appropriateness of the options employed.
12. Steps should be taken to analyse the sustainability of green productivity programme.

REFERENCES

Asian Productivity Organization (2000). Green Productivity in Asia: APO's Demonstration Projects 1995-1999. Asian Productivity Organization.

Cascio, J. (ed.). (1996). The ISO 14000 Handbook. Milwaukee: ASQ Quality Press.

Culaba, A.B. and M.R.I. Purvis. (1999). A Methodology for the Life Cycle and Sustainability Analysis of Manufacturing Processes. *Journal of Cleaner Production*. 7: 443 - 445.

De Simone, L.D. and F. Popoff. (1997). Eco-efficiency: The Business Link to Sustainable Development. World Business Council for Sustainable Development.

Foust T.D. and D.D. Gish. (1996). Future Perspective. In: Curran, M. A., (ed). *Environmental Life-Cycle Assessment*. New York: McGraw Hill.

Giarratano, J. and G. Riley (1998). *Expert Systems: Principles and Programming*. Boston: PWS Publishing Company.

Heijungs, R., (ed): (1992). Environmental Life-Cycle Assessment of Products: Guide—October 1992. Report 9266. CML (Centre of Environmental Science) in Leiden, TNO (Netherlands Organisation for Applied Scientific Research) in Rotterdam.

Ong, Victoria D. (1997). Design and Development of a Microcomputer-based Productivity Diagnostic. Prototype Utilizing Embedded Rule-based Expert System. MS Thesis. University of the Philippines.

Porter, M.E. and C. van der Linde (1995). Omen and Competitive. *Harvard Business Review*. 120, 122, Sept-Oct 1995.

The Netherlands ISO (1997). ISO 14040: Environmental Management-Life Cycle Assessment - Principles and Framework. International Organization for Standardization.

UNEP (1993). Environmental Management in the Electronics Industry: Semiconductor Manufacture and Assembly. *Technical Report* No. 23.

Watanabe, S. (1993). *Resource Productivity as a New Measure for Industrial Performance*. Sony.

WCED (1987) Our Common Future, World Commission on Environment and Development. Oxford University Press.

9

INFLUENCE OF BRANDS IN PACKAGED DRINKING WATER INDUSTRY

T. Karthikeyan and *V. Balakrishnan*

Introduction

The brand is increasingly becoming the key source of differentiation that guides customer purchase choice. It is the focal point around which an organization defines how it will uniquely be delivery value to the customer. If the brand is well conceived and consistently delivered through all business processes and customer contacts, the organization will grow and prosper. Not too long ago, marketers at consumer product companies seemed to be the only ones interested in talking about brand management and branding. But these days all kinds of organization are recognizing the importance of branding. In the past few years museums, universities, restaurants, churches and individuals too build their brands. However, the average person on the street is talking about brands.

Philip Kotler[1], views branding as a major issue in product strategy. Brands as a complex symbols convey up to six levels of meaning. They are attributes, benefits, values, culture, personality and user. ICRA[2] views the term fast moving consumer goods although popular and frequently used, does not have a standard definition for that purpose the term has been defined to include products that are consumed atleast once a month, fold under national brands in consumer packages through a widespread distribution network and consumed directly by the consumer.

The FMCG business in India is characterized by low capital intensity, stable and easily available technology, high initial launch cost, relatively stable volume of reward and a large and diverse consumer base. Brands play a vital role in this sector.

Under the Foods (Product Category) the products include chocolates, confectionary, snacks, cooking media, beverages, dairy and processed milk products. In Beverages category the packaged drinking water industry seriously affects a Nations Safety, Health and Environmental Sectors.

Objectives

This paper discusses the different environmental issues caused by brands of packaged drinking water industry. The paper has the following objectives :

1. To examine the role of brands in packaged drinking water industry.
2. To discuss the safety, health and enviromnental related issues caused by package drinking water brands.

Indian Packaged Drinking Water Industry

In the year 2002, packaged drinking water industry in India was a Rs. 1000 crore industry. The packaged drinking water industry grow by 40 per cent every year. The product differentiation is considered to be critical factor among the major competitors in the market. Major competition like Coke, Pepsi, Bisleri and other major players hold the entire packaged drinking water industry, where the Coca Cola's Kinley brand faced stiff competition from Ramesh Chauhan's Bisleri and Pepsi's Aquafina. The Indian market for packaged drinking water market shares in July (2002) is portrayed in the Fig. 9.1.

Role of Brands in Packaged Drinking Water Industry

Human needs includes food, shelter and water. But now water even though considered to be an essential product it cannot be consumed

Fig. 9.1 : Indian Market Packaged Drinking water Market Share (2002)

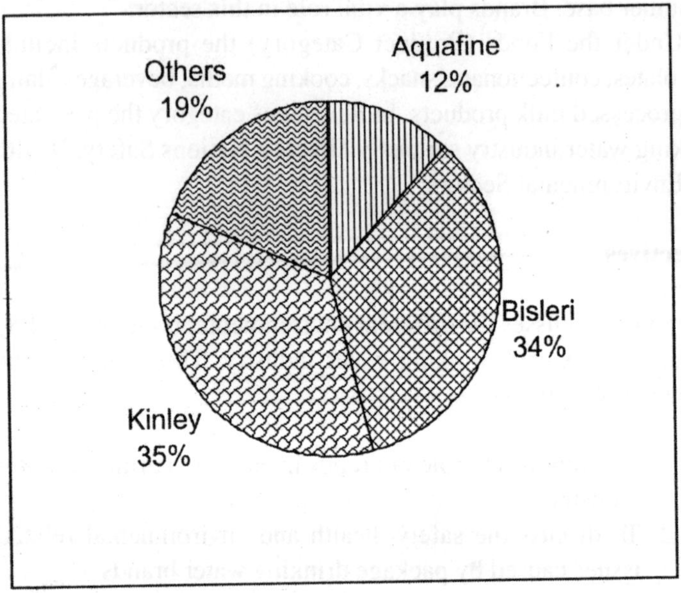

Source : ORG–Marg AC Nielson (2002).

by every customer in the market. Oil rich countries earlier used to buy and feel the scarcity of water but now in India customers are forced to buy packaged drinking water. Customers equity towards Pepsi is reflected over Aquafina, Coke's equity is reflected over Kinley and Bisleri as such. Bisleri was the first brand to get introduced to the concept of packaged drinking water industry. Even now customers in the market ask for Bisleri (meaning that mineral water) in any petty shop. Ramesh Chauhan estimated in the year 2000, that Bisleri company would grow at the rate of 40 per cent. So that Bisleri will become a Rs. 1000 crore brand in future.[3]

Factors—Packaged Drinking Water Industry (PDWI)

Population growth, LPG effect, political scenario, growth of tourism industry, hotel industry, hospitals, changes in lifestyles of consumers, availability of products all over the country, are the major factors determining the growth of packaged drinking water industry.

Safety and Health Issues of Packaged Drinking Water Industry (PDWI)

The issues related to safety and health aspects of Packaged Drinking Water Industry (PDWI) is discussed seriously since contaminated water is supplied to customers to be safer side the brand owners do use the words as purified safe drinking water in the wrappers of the bottles.

Buy Water, Drink Bacteria

The pollution monitoring laboratory of the Centre for Science and Environment (CSE), Delhi has found that most mineral water brands in India contain residue of killer pesticides. Scientists in (CSE) tested 17 major brands in Delhi and 13 brands in Mumbai and found that unacceptable levels of lindane, DDT, Malathion and Chlorpyrifos were present in played drinking water industry[4].

The raw water from their bottling plants were also tested. These samples were subject to the internationally accepted tests approved by the US environment protection agency. Their results showed a deadly cocktail of pesticides. High level of organo phosphorous and organo chloride compounds were also present in the sample. The Malathion and Chlorpyrifos can cause neuro behavioural changes, vibral difficulties and even autism in children. Lindane is capable of causing asthma and liver malfunctioning.

The lab test as samples from Delhi showed 36.4 times more pesticides than permissible levels while sample from Mumbai brands were 7.4 times more than permissible levels. The top selling brands like—

- Bisleri showed pesticide concentration levels 79 times higher.

- Kinley 14.6 times higher than the norm.
- Aquaplus topped the dubious list, crossing the limits by 104 times.

Evian, a brand that bottles water at source in the Alps in France, is the only brand that did not contain any pesticide. But the Evian is beyond the reach of the average Indian.

The source water for these bottling units were from bore wells in industrial areas which results in high level of pesticide presence. There are no regulations in India to ensure that bottled water plants are to be set up in clean ground water zones.

Environmental Issues in Packaged Drinking Water Industry (PDWI)

Issues like wells running down dry, scarcity of groundwater, contamination of groundwater need to be addressed.

Wells Run Dry?

In Kerala, Plachimada, Palghat district where cokes Greendfield soft drinks bottling factory was situated, the wells run dry due to drawing of 1.5 million litres/day from the common groundwater resource. In the year (2002) the Coke could extract only 800,000 litres due to water scarcity in the groundwater resource. The remaining water is supplied by water trucks that carry water extracted from borewell in neighbouring villages. The exploitation of water resource leads to excessive calcium and magnesium in ground water.

Groundwater Scarcity?

The construction of bottling plant by Shakthi sugars and Coca Cola combine in Padamathur in Sivagangai Tamil Nadu drastically affected the level of groundwater. The groundwater level in Padamathur has been going down everyday since the establishment of the Shakthi sugars. The two pumping stations installed by Shakthi sugars-Coca Cola continue in the Vaigai river bed in Kanoor village which extracts 10 to 12 lakh litres of water per day is the major cause for the unprecedented groundwater drought leading to the drying up of irrigation wells and drinking water source within the radius of 7 km from the pumping station unrestricted extraction of groundwater by Shakthi sugar-Coca Cola combined is depriving the farmers, women and the general public of their legitimate right on ground waters.[5]

Measures to Improve Packaged Drinking Water Industry (PDWI)

1. Water has to be graded weekly on various factors like purity, softness, cleanliness etc.
2. High level of Brand knowledge has to be built among consumers towards Packaged Drinking Water Industry.
3. More green movements have to develop to care country's natural resources.
4. Government must enact ground water policy according to priority to each sector like Drinking Water, Agriculture, Feeding of cattle and livestock, Water for social sector, and water for industrial and commercial purpose.

Conclusion

In Chennai alone more than 200 legal and 400 illegal bottled water manufacturers operate Indian PDWI which is barely a decade old is already Rs. 8 billion ($ 170 million) market and is expected to grow by Rs. 12 billion ($ 250 million) by 2004. Brand names such as Cokes, Kinely, Pepsi's Aquafina and the Indian bottled water giant Parle's Bisleri enjoy Rs. 4 billion ($ 85 million) market. The profitability of their business is beyond question.

Dhanal G. Shah[6], opined that the green movement has been picking speed in the past decade changing people's attitude towards consumerism. The demand for environmentally, friendly and ethically correct product is increasing. A new breed of tourist is emerging known as green tourists. These green tourists, travelling independently, or in small groups, living as local and caring for the country's resources. This definitely is the pollution age but also the era of green areas.

NOTES

1. Kotler (2003), *Marketing Management*, Prentice Hall of India.
2. ICRA (2001), *Industry Watch Series Fast Moving Consumer Goods.*
3. *Indian Express* Newspaper (Bombay) Ltd., (2000), "Ramesh Chauhan to make Bisleri Rs. 1000 crores brand," March 31.
4. Panicker, Ajay (2003), "Deadly Cocktail", *The Week,* Feb. 23.
5. Jayaraman, Nityanand (2002), "No Water Drink Coke", India Resource Centre, May 28.
6. Dhanal G. Shah (2003), "Eco-tourism", Advanc'edge MBA/August.

10

SOLUTION FOR POLLUTION FREE PRODUCTION BY KVI SECTOR

S. Radhakrishnan and *K. Balasubramani*

Introduction

Khadi and Village Industries (KVI) Sector process locally available raw materials and thereby giving indirect employment to the agricultural labour. During the off season it provides enployment to landless agricultural labour. Supply the society with quality products and take care of the health of the people. It is not using any chemical or sophisticated machinery etc. It is not polluting the environment (this is discussed separately).

As the people get employment, food etc. they don't find time to indulge in any other unlawful activities. Hence Khadi and Village Industries Sector, both directly and indirectly facilitating to have peaceful social life in the society.

Benefits to the Agricultural Sector

Khadi and Village Industries Sector consumes agro-based raw materials and thereby helping the farmers to grow more agricultural products. It also processes forest based oil seeds, and other materials. Bee-keeping industry of Khadi and Village Industries Sector helps the agricultural sector for better production by cross pollination. It supply neem oil cake, to the land as fertilizer and neem oil as pesticides, and also supplies groundnut oil cake to the animals maintained by the farmers. Khadi and Village Industries and Agricultural Sector are closely linked with each other and doing better service to the society.

Benefits to the Government

There are three 'E's which may often creating problems to the Government.

1. Energy
2. Employment
3. Environment

Any energy is saved is equal to energy produced. Khadi and Village Industries Sector hardly uses power, or mineral oil for the purpose of production. It uses mainly simple tools and machinery. Major production is by the masses only and not by huge machines. Hence this sector saves so much energy and help the Government to get rid of energy crisis.

Next major problem is unemployment especially rural employment to the unskilled rural people. Khadi and Village Industries Sector alone can provide employment to the unskilled rural masses, with low per capita investment. Large scale and corporate sector genetated below one per cent enployment with huge investment. Khadi and Village Industries Sector generated 56.50 lakhs employment with the per capita investment of Rs.20,000. This sector helps the Government, by providing employment to rural mass and reducing the problems of unemployment. The major problems due to unemployment is anti-social activities. Idle hands are used for the anti-social avtivities by the cheap, selfish politicians which may create lot of social problems. To have peaceful society the role of Khadi and Village Industries Sector is very much essential.

The third 'E' is environment. This draws attention from all over the world. Any nation violating the environmental laws becomes not eligible to receive aids from foreign countries expecially from USA, based world orgnisations like World Health Organisation (WHO), UNICEF etc. The problems of environment and the role of KVI Sector in promoting environment is discussed below.

Benefits to the Environment and Eco-system

Ancient Indian Sages quoted that in the current age of Kaliyuga, the sins of mankind are being so bad that the Earth and mankind would be destroyed in the near future. What these sages said may not have been a mere fantasy considering the environmental pollution which the people have inflicted on this planet, thereby contributing to the destruction of life on the Earth. Continued economic growth, mismanagement of resources and population explosion have an explosive impact on the environment. Now our Earth has become a very sick planet and urgently needs a cure. A disaster is looming unless everybody joins hand to take on the environmental dangers and to find ways to protect society from half the march to mankind destruction.

Environmental Pollution

Pollution is an undesirable change in the physical, chemical or biological characteristics of our air, land and water that will harmfully affect the human life and the desirable species or they may waste of deteriorate our raw material resources.

Pollutants

Any substance, which causes pollution is called a pollutant. Pollutants mean "any solid, liquid or gaseous substance present in such concentration as may be tend to be injurious to the environment".

Pollutants are the residues of things we make, use and throw away. There are many sources of such pollutants. The lakes and rivers are polluted by wastes from chemical and other factories, and the air, by gases of automatic exhausts, industiries, thermal power plant etc.[2]

Environmental Pollutants

The various principal pollutants, which pollute our air, water, and land are as follows :

1. Deposited matter— soot, smoke, tar, dust grit, etc.
2. Gases—oxides of nitrogen (NO, NO_2), Sulphur (SO_2), Carbon Monoxide, Halogens (Chlorine, Bromine, Iodine).
3. Acids Droplets—Sulphuric acid, Nitric Acid etc.
4. Fluorides.
5. Metals—Mercury, lead, iron, zinc, nickel, tin, cadmium, chromium, etc.
6. Agrochemicals—Pesticides, herbicides, fungicides, nematicides, bactericides, weedicides and fertilisers.
7. Complex Organic Substances—benzene, ether, acetic acid, benzpyrenes etc.
8. Photochemical oxidants—Photochemical smog, ozone, peroxyacety nitrate (PAN), Peroxybenzoil Nitrate (PB_2N), nitrogen oxides, aldehydes, ethylene etc.
9. Solid Wastes.
10. Radioactive waste.
11. Noise.[3]

Causes for Pollution

The following are the main causes of pollutions:

Human activity: The main cause of pollution is man himself. His activities, civilization, culture etc. Bring about environmental pollution. Hence pollution is said to be a 'necessary evil' in the percentage.

Population Explosion: It is stated that population explosion causes pollution because more people produce more sewage, more solid wastes, more fuel being burned, more fertilizers and insecticides used to produce more food.

Industries: Paper mill, Sugar factories, Soap factories, Cement factories, Chemical factories, Fertilizer factories, Oil plants, Rubber factories, Blast furnaces, Distilleries, Oil refineries, Leather industry etc. cause pollution to a greater extent.

Automobiles: Automobiles like road vechicles, trains,

aircrafts etc. causes air pollution and noise pollution.
Smoke: Smoke arises from industries and houses.
Biocides: Insecticides, pesticides and other biocides cause pollution.
Fertilizers: The synthetic fertilizer used for the improvement of crops cause pollution.
Sewage: It is the liquid pollution emerging from houses and industries. The domestic sewage contains urine, faecal matters, kitchen wastes, etc.[4]

Types of Pollution

Pollution is classified into following types :

Air Pollution,
Water Pollution,
Land Pollution,
Noise Pollution, and
Thermal Pollution.

Industry-wise Pollution

Common Industry Wastages
Pollutant

Cadmium, Zinc, Chromium, Polythene pack etc.

Pollution

Water Pollution

Physico-chemical characteristics of the river Vaigai water samples, influenced by sewage average like pH value 7.9. Temperature of the water ranges from 28°C to 32°C. Total suspended solids (mg) 0.13 to 2.90, dissolved oxygen (mg/1) ranges from 1.30 to 5.6 biological oxygen demand (mg/1) ranges from 0.52 to 2.87, metals of pH ranges from 0.64 to 2.94, zinc 0.13 to 1.50, Cadmium ranges from 0.18 to 1.101.[5]

Land Pollution

Similarly the quality of soil is also affected enormously. That is the alkalinity of the soil is reduced to high level, which leads to the low quality of soil.

Affected to Man

Such type of waters cannot be used for drinking purpose. Where the people drink the above type of river water. which affect their biological system.

Paper Industry

Land application of wastewater is a preferred alternative for its disposal since solid is believed to have a capacity for receiving and decomposing wastes and pollutants, where organic materials are stabilized through the activity of micro flora in the soil.[6] Among the major industry in India pulp and paper is one of those that contribute significantly to water pollution. The quality of fresh water used in the production of 20 tonnes of paper is about 8000m[7] per day. Nearly 75 per cent to 95 per cent of fresh water used in the paper and pulp mill was discharged as effluent containing organic and inorganic pollutants and colouring materials. Effluents are used for irrigation in dry land areas after treatment. These effluents not only contain nutrients that enhances the growth of crop plants also have toxic materials effect of paper mill effluent under field situation like psycho-chemical characteristics, biological properties and enzyme activities.

Physio-Chemical

Tamil Nadu Paper Limited (TNPL) Sludge added plots recorded higher pH values and compared to well water irrigated plots, effluents alone received treatments recorded higher pH value like 8.41, organic mater 0.73 and bulk density 1.33. In addition to that the hardness, electrical conductivity, Total Dissolved Solids (TDS), Chemical Oxygen Demand (COD), BOD and other toxic materials like cadmium. Lead etc. are also varied. When the above parameters vary from their tolerance limit, the quality of water changes.

Biological Properties Soil Rhizospher

The available nitrogen, phosphorus and potassium, status was higher in the effluent plots than the well water irrigated plots. The available nitrogen content carried from 180.60 to 205.80 and 161.07 to 182.44 kg ha^{-1} in the effluent and well water irrigated soils respectively. The available phosphorus content varied from 10.54 to 15.24 kg ha^{-1} in effluent irrigated plots and 9.39 to 11.69 $kgha^{-1}$ in well water irrigated plots. Thus the effluent is more useful for irrigation than the well water. So, the plants growth and development are well pronounced.

Soil Enzyme Activities

The Soil amylase activity increased in effluent irrigated plots. All the treatments showed significant variations in the main as well as interaction effects. The amylase activity varied from 0.35 to 0,61 and 0.29 to 0.46 mg in the effluent and well water irrigated plots respectively. The activities of inverts was higher in amendment received plots than in the amended plots. The cellulose activity of the effluent irrigated plots varied from 12.04 to 17.25 mg but it varied from 10.6 mg in well water irrigated plots. The CO_2 evolved showed a variation of 3.44 to 5.16 mg in effluent irrigated plots and 2.63 to 3.58 mg in well water irrigated plots. The fertility of soil is associated with the activity of amylase, inverts and cellulose. When the industrial effluent is combined with microbial-load, the soil fertility had to be enhanced. The mechanism in the soil are involved in various decomposition and chemical transformation in the soil.

The above paragraph revealed that irrigation with treated effluent increase the rhizospher microflora soil enzyme activities.

The above study revealed that pulp and paper mill effluent irrigation to soil was not harmful to the soil microbial population and its activities.[8]

Solution

Hand made paper industry (HMP) is one of the major solution to the above problem of small paper industry pollution. Because hand made paper industry using the low level of chemical substance, but, the raw materials in the Hand-made paper industry are waste paper, rags, sun hemp, banana fibre, jute, husk, and other waste material. While mill made paper industry using the major raw material as bamboo, wooden scrap, paper pulp etc. Thus the hand made paper industry is one of the alternative industries for pollution control and save the environmental properties (like save tree).

Tannery Waste

The wastes from this industry rank among the most polluting of all industrial wastes. It is probably the first industrial waste problem of mankind[9]. Tannery effluent, being voluminous and highly putrefactive in nature, when discharged untrapped, damages the eco-system of the receiving steam. On the basis of process and chemicals used during the tanning operation, sulphides, chlorides chromium, insecticides, arsenic, high pH, high Biological Oxygen Demand (BOD), suspended and dissolved solids constitute as major pollution in the effluent. It is interesting to note that the total contribution from tanneries of BOD load on river Ganga of Kanpur is 20 per cent which total volume is only about 1 per cent.

Differing views have been expressed on the effect of chromium bearing tannery effluent add sludges on plant life and soil productivity. Chromium occurs in liquid wastes in two different forms, trivalent and hexavalent. Hexavalent chromium is toxic and known to be a carcinogenic substance. It is responsible for lung cancer, chrome ulcer, perforation of nasal septum and kidney damage.[10] The major polluter of the tannery effluent is chromium. In plants, the rates of plants accumulated large amount of chromium.

Than the leaves and flowers. The translocation of chromium in plants mainly in the form of hexavalent it would have facilitated larger accumulation in the roots. The experimental data explained that hexavalent loans don't have any toxic symptoms in flowers and crops.

Land

There was a marked impact of tanneries effluent and soil properties. That is the soil pH and EC where decreased initially and than increase after some days. There was not much change in fertility of the soil but a small increase in Nitrogen, Phosphorus and Potassium (NPK). The translocation and reduction of chromium (+6) to (+3) favours least accumulation of chromium in plants. This is, the quality of ground water is damaged by pollutant chromium through the process 'leaching'.

Air

Gases affect the quality of air in the atmosphere. *Physico-chemical analysis of tannery waste.* The physical chemical characteristics of tannery waste and studied. The parameters pH, Electrical Conductivity (EC), Carbonate, bi-carbonates, sodium, calcium of tannery effluent are studied. Such effluent affect the metabolic process of plant. The harmful gases like SO_2 (sulphurdioxide) affects the micronutrients such as zinc, copper and manganese. Thus the yield of crop is affected.

To obtain air pollution, effluents from tannery industries are highly harmful to the human society. So of prevention the pollutants and effluents is more important. The following suggestions are very useful for preventing thc above factors to some extent.

In KVI Sector for tanning the raw hides, natural herbal products and lime are used, which will not create and effluent problem.

Fertilizer Factory

Fertilizer factory are polluted the water, land pollution. For example the Damodar was studied with respect to the effluent discharged from the fertilizer factory at Sindri. The Damodar river runs through the coal belt of Bihar encompassing Sindri. Damodar are seasonal rain fed river passing through this industrial belt. Both of them are utilized for generating hydro-electric power. The discharge of effluent from the fertilizer factory into the Damodar river occurs via this Sindri nalla and aggravates the polluted condition already brought about by a load of suspended solids in the form of coal dust and fly ash released in the river from coal washeries, located along its length. The daily discharge of 18,000 M^3 of wastes from the factory includes alkalis, chromated ammonia, cyanide and phenols. Biology oxygen demand level is high due to the chemical released in the effluents level is high. Added to this the high oxygen demand of the pollution resistant species which utilized the available nitrate and phosphates to increase their concentration considerably. Similarly, the presence in water of reluctance such as ammonia, nitrates readily oxidable substance, increase the oxygen demand and develops an oxygen deficit, producing on anaerobic conditions in which the microbes thrive. Thus, the anaerobic condition, high concentration of Total Suspended Solids (TSS) nitrogenous substances, acidic media, high ammonia concentration presence of poison such as cyanides and phenols have completed wiped out the normal fresh flora and fauna from the part of the Damodar river under study. In this high pollution zone, only pollution tolerant specles are dominant and their in anaerobic conditions with moderate nutrients.[11]

The chemical fertiliser industries are also important part explaining the quality of soils and drinking waters. In general, the chemical fertilizer industries are located at river belt. The discharged chemical effluents are damaging the aquatic life, nutrients of waters and other living organisations. Usually, the effluents contain highly poisonous substances. Thus, simple and useful way of preventing the chemical effluents from fertiliser industries are the adoptation of bio-fertiliser. Because bio-fertilisers don't degrading the quality of drinking water and also enhance the growth of micro living organism in water.

It is proved beyond doubt that lot of social benefits as discussed above are accrued from KVI Sector, but earn meagre net profit which is not worth noted.

NOTES

1. Khadi and Village Industries Commission: "Statistical Abstract in Khadi and Village Industries", KVIC, Mumbai, 1998.
2. Arumugam, N., : *Concepts of Ecology (Environmental Biology)*, Saras Publication, Nagercoil, p. 214.
3. Sharma, P.D. : *Environmental Biology*, Rastogi Publications, Meerut, 1999, p. 155.
4. Arumugam, N., *op. cit.,* pp.215-216.
5. Kandhasamy, M. and Santhaguru, K, : "Influence of Sewage on Physico-chemical Charateristics of the River Vaigal", *Journal of Ecobiol,* 6(4) 1994, pp. 315-217.
6. Young, J.C., McDermoff, G.N, Jenins, D. 1981 : "Alterations in the BOD procedures for the 15th edition of Standard methods for the exanination of water and waste water", *J. Water Pollut.Control Fed.* 53: 1253-1259.
7. Sastry, C.A., Kothandaraman, V. and Aboo, K.M., 1977: "Treatment of Waste Water from Small Paper Mill without Sada Recovery—A Case Study", *Indian J: Environ.* H, 19: 346-359.
8. Dhevagi P., Rajannan G. and Oblisami G,: "Effect of Papermill Effluent on Soil Microflore of Maize", *Journal of Indus. Poll. Control* 16 (1) (2000), Pp. 95-105.
9. Pandey, G.N. and Carnety, G.E. : *Environment Engineering*, Tata McGraw Hill Publishing Company Ltd., New Delhi, 1989.
10. Krishnaswamy, N.R., Gopalakrishna, K. and Sheela Evangeline, Y. "Effect of Chrome Tanning effluent on cohesive soil," *Journal of Indian Water Work Association*, April-June.
11. Singh, S.N. : "Effluent from the Sindri Fertilizer Factory in the River Damodar", *J.Ecobiol*, 6(1), 1994, pp. 027 to 032.

11

SOLID WASTE MANAGEMENT TECHNIQUES IN RURAL INDUSTRIAL UNITS – A MICRO ANALYSIS

N. Murugan and *M. Sivakumar*

Introduction

The importance of Small Scale Industries (SSIs) cannot be underplayed in the industrial development of any country. It continues to be an important part of the informal sector in India as a generator of employment and income. The SSIs have, therefore, been the focus of the industrialization strategy and have received special emphasis in successive five-year plans in India. Many small and medium industries are also generators of hazardous wastes like heavy metal sludges, solvents, waste oils, acidic and alkaline wastes, photo wastes, etc. They use unsafe method of disposing the industrial wastes, e.g. dumping solvents or heavy metal laden waste water into storm water drains or in a commonplace, which are notoriously difficult to monitor. Recent years have seen a growing concern about the impact of such pollution on environment in India and hence, the government is focusing on special measures and innovative approaches to address these pollution problems. Until recently, the implementation of environmental laws has been the exclusive province of specially constituted state agencies, which issued permits and prosecuted violators in the ordinary criminal courts. Since 1986, enforcement techniques have been strengthened. These laws now permit precautions at industry level and more significantly, empower state agencies to shut down polluting industries or stop their supply of water and power.

Solid waste management in India is an emerging and engaging area of study. However, the picture is often confusing and solutions fuzzy as information available in public domain are either scanty or scattered. This paper attempts to discuss the solid waste management practices in rural industrial sectors.

Solid Waste Management

Higher standards of living of ever increasing population has resulted in an increase in the quantity and variety of waste generated. It is now realized that if waste generation continues indiscriminately then very soon it would be beyond rectification. Management of solid waste has, therefore, become very important in order to minimize the adverse effects of solid wastes.

The solid waste can be managed through " Three R's", i.e. Reduce, Reuse and Recycle.

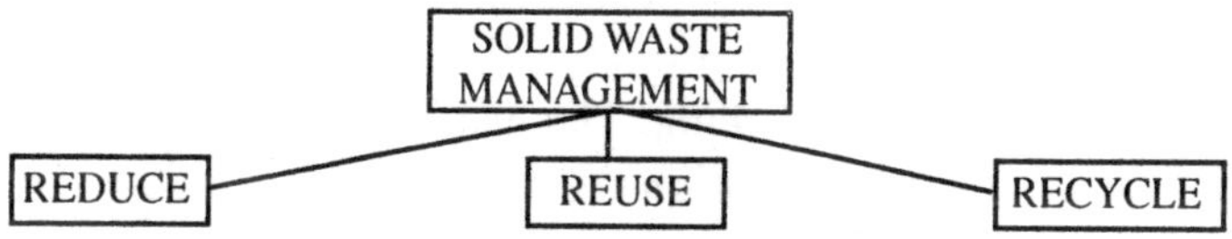

Classification of Solid Wastes

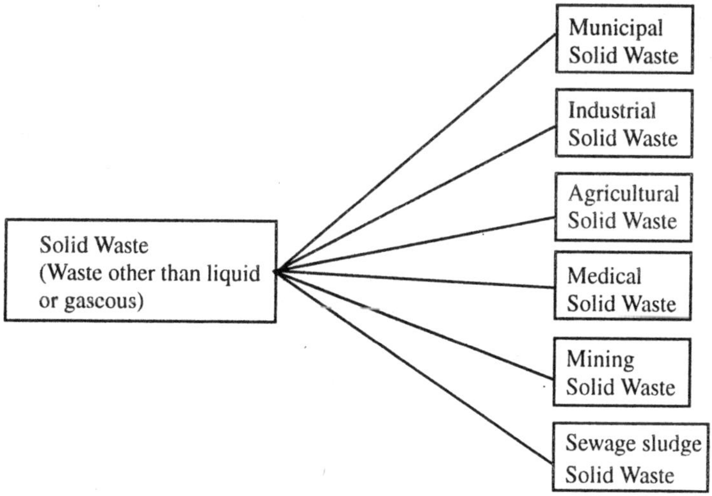

Rural Solid Wastes

There are various types of wastes in rural areas, namely community wastes, wastes from agricultural and agro-based industries, animal wastes and oil bearing seeds etc. Table 11.1 provides estimated annual generation of various types of rural wastes in India (Adapted from Report, 1990). The community wastes from rural area are estimated at 5 million tonnes of night soil and 10 million tonnes of refuse. The rural population of 629 million (Census, 1991) is distributed over nearly half a million villages. This makes for a small population per rural settlement. Additionally, the population densities are also very low compared to highly urbanized areas. Due to these reasons, collection and transportation of rural wastes in India is not a pressing problem. Low overall volumes also do not necessitate institutional structures for its management. For using as fuel, animal dung is shaped into cakes and dried and stored to be used for domestic cooking. The excess is also used to make compost for farm applications. Composting is carried out by accumulating dung, domestic and other wastes in a heap or pit.

Agricultural residues are largely used as animal feed; a small portion is also used as fuel and as construction material. Per cent utilization of rural wastes for various end uses is outlined in Table 11.2 (Adapted from Report, 1990). It is found that traditional practices of using wastes by way of fuel, animal feed and farm manure accounts for nearby 90 per cent of all waste utilization. Barely 1.6 per cent of wastes are not being utilized for any useful

Table 11.1. Estimated Annual Generation of Various Rural Wastes in India

Types of Waste	*Estimated Percentage*	*Generation of Total Waste (Million tonnes)*
Community	15	0.81
(a) Night soil	5	-
(b) Refuse	10	-
Agricultural residues	322	17.4
Animal dung	1365	73.74
Agro-industrial by-products	49	2.65
Oil seeds	100	5.4

Table 11.2. Percentage Utilization of Rural Wastes for various End Uses

End-use	*Percentage Utilization*
Fuel	14.8
Animal Feed	35.0
Farm Manure	39.3
Construction Material	2.8
Industrial Raw Material	1.5
Other Uses	4.9
Waste	1.6

purpose. It is clear that traditional methods have been adequate in handling wastes generated. Hence, rural solid wastes do not constitute a problem area like urban solid wastes. However, a case for S & T inputs does exist. Technologies such as improved *Chulhas* (wood/dung stoves), Bio-gas from night soil or agro residues, Biomass Densification, Gasification and Pyrolysis offer a way to enhance energy efficiencies and ensure efficient utilization of available resources, in addition to improving quality of life for rural masses.

The use of certain wastes as industrial raw material is limited to only 1.5 per cent. However, a number of possibilities exist to ensure increased industrial utilization of rural wastes for making value-added products. Rice husk based particle boards, Bagasse based paper, and charcoal, and packaging material, chemicals through bio-conversion and industrial fuel are some examples.

Urban Solid Waste

With industrial progress, growing urban areas and resultant growth in urban solid wastes is a relatively new phenomenon in contemporary India. During mid-seventies, the per capita solid waste generation ranged from 150-350 gm/day for various Indian cities (Bhide *et al.* 1975), whereas in late eighties, it ranged from 320-530 gm/day. The urban population is currently about one-fourth of total population. It is projected to be nearly one-third by end of the century. The total urban population in 2001 is estimated to be around 330 million from current 218 million. The class I towns alone account for nearly 60 per cent of all urban population (Census, 1991). The traditional knowledge systems primarily evolved for rural and dispersed populations have not coped up well with densified living conditions and associated need for basic infrastructure and its management. Relatively poor management of urban wastes is reflected in degradation of living environment of urban areas. The average per capita solid waste generation and collection efficiencies data for various categories of towns in India. Whereas, the large towns show distinctly higher per capita waste generation; there is no significant difference between medium and small towns. The collection efficiencies in small and medium towns lag far behind at nearly 60 per cent than those of large towns at > 80 per cent. These points to weak infrastructure and poor financial status of small and medium towns. The physical characteristics of Indian urban garbage. The figures presented are on wet basis and the moisture levels can range from 40-70 per cent.

The wastes largely comprise of bio-degradable organics. High moisture and organic content coupled with high prevailing temperatures make frequent removals necessary. This places additional burden on already overstrained system. The collection of refuse presents peculiar problems as household wastes are thrown out indiscriminately. Also due to narrow lanes, collection vehicles can reach only selected accessible points. Hence, unskilled labour is used to sweep streets and collect garbage. Though labour rates are cheap but due to large scale manpower deployment and low productivity, the costs are high. It is estimated that India spends four times as much on sweeping as on refuse collection (Pickford, 1983). Poor motivation of workers, inadequacy of supervisory and management skills at local government levels are other leading causes of low productivity. The problem needs attention at appropriate levels. The cost of collection in India tends to be a very large part of overall solid waste budget. To cite an example, the city of Ahmedabad with 3 million population and 1260 tonnes of solid waste per day, spends 85.8 per cent of its budget on collection, 13.4 per cent on transportation and only 0.8 per cent on final disposal (Report, 1990). The benefits from present level of expenditure can be enhanced by following better methods of collection, efficient transportation; appropriate technology induction, better management practices and motivation of workers. The three R's of waste management, namely Reduce, Recycle and Recover are oft-repeated phrases in Indian policy circles. However, what is lost sight of is that culturally there is no propensity to waste. Also, there is a thriving informal sector of recycling. This recycling is achieved through *Kabaris* the waste handlers, who go from door to door and collect used bottles, broken plastics, metals, waste paper etc. This material is then traded for manufacture of secondary products for which markets exit.

Scavengers foraging through wastes are an unhygienic practice. It still, however, contributes to recycling effort in India. The scavengers act as the second filter after *Kabaris* have taken away first batch of useful materials for secondary market. The income from foraging provides much needed subsistence to poorest of the urban poor. A ban on scavenging on health grounds may seem like a solution but will only aggravate the problem of basic sustenance However, providing facilities such as bathing at the end of days work to foragers will go a long way towards improving their health status. This market driven mechanism of segregation at source is a positive feature of Indian urban waste scenario. There are no figures available to estimate the volume of waste being processed, as this sector is not documented. The secondary product market though strong is not regulated. The specifications and quality is often sacrificed, e.g. a lot of mixed plastic waste is reprocessed to make containers etc. These being cheap are bought readily by the consumers. However, storage of food stuff in these containers can be harmful for human health. There is a definite need to examine secondary products and to regulate effectively to safeguard human health.

Sources of Urban and Industrial Wastes

Urban waste consists of medical waste from hospitals; municipal solid wastes from homes, offices, markets (commercial waste) small cottage units, and horticulture waste from parks, gardens, orchards etc.

Land Filling

Like other Asian countries, in India too most of the waste is landfilled. The methods followed are not in keeping with modern practices of sanitary landfilling. The wastes are largely dumped. This dumping is normally carried out in low lying areas which are prone to flooding. During rainy season, possibility of surface water contamination increases due to flooding of these low lying areas. The ground water pollution though largely unassessed is another threat posed by dumping of wastes.

The daily cover techniques are poor leading to vector problems. The birds foraging on garbage dumps are known to cause substantial problems for aircrafts operating in the urban areas. The bird strikes have resulted in a great deal of loss to aviation sector. This state of affairs results from lack of knowledge and skills on part of local authorities. Diversion of large part of money to collection and transportation of wastes results in non-availability of funds for disposal activities. This forces local authorities to curtail even known precautions and practices and use short cut approach.

Composting

Composting is a highly suitable option for urban solid wastes in India. High organic content and moisture make it particularly attractive. Conceptually, the idea of composting is appealing as it helps to recycle the nutrients back to land. The process, however, requires segregation of inert material; which is achieved easily due to recycling by *Kabaris* and scavengers. This option, hence, appeared ideal in mid-seventies when a number of compost plants were set up in various cities. Mechanised aerobic composting offered hope for big towns starved of landfill space. The plants were commissioned during 1977-80 and were operated either by State Agro-Industries Corporations or by Municipal Corporations.

These plants were expected to provide much awaited answer to growing problem of urban solid wastes but operational and other problems began to appear. Due to low skill/managerial inputs the operating efficiencies were low resulting in high cost of production. The problem was further compounded due to large distances between compost production centres and the compost utilization centres, namely the farmlands. The resulting cost of transportation made marketing even more uneconomical. Farmers also reported problems with broken glass pieces in the compost.

A good way to avoid these problems is to intercept suitable wastes at the point of generation before it is mixed with other wastes. Kitchen and vegetable market wastes are largely suited for this purpose. These wastes can be collected and treated at source, if space permits. The resulting bio-gas can be used for captive energy use such as lighting and cooking etc. Few Bio-gas systems are currently available to treat wastes of fruit and vegetable origin (Nagori *et al.* 1988). Though currently unfeasible as a large scale option, Bio-gas systems can effectively handle localized and specific wastes and contribute to environment friendly disposal of wastes.

Refuse Derived Fuel (RDF)

This method of waste disposal primarily views waste as a resource. After separation and size reduction, the combustibles can be palletized. Integrated Waste Management project at Bombay attempted to do just that. Due to local conditions, the product off-take and price realization was estimated to be good. This avoided the earlier problem faced by composting plants. The large scale processing of garbage was also supposed to considerably slow down exhaustion of landfill space in the near vicinity of the city; obviating need to spend much larger amounts on transportation costs. This pilot technology development effort also offered prospect of totally indigenous and cheap technology. The cost of 80 TPD plant was Rs. 15 million or nearly half million US dollars (May 94). This compared very favourably to New Delhi incineration plant (300

TPD, Rs. 220 million). As it was first attempt of its kind, it required experimentation and modifications to zero down on specific waste handling, size reduction and separation processes along with optimization of system parameters. The plant was erected and extended trials were undertaken. A number of new innovations were made in garbage separation methods. The fuel pellets produced were also test marketed successfully. However, there was a need to support the technology development effort for a long enough duration which has been lacking. Despite the promise of RDF, it will be limited in application due to need to have large industrial areas in close proximity to market the fuel to. The cost differential between cost of coal and the RDF should also be attractive to ensure sales.

The composting, however, still remains a strong option for small and medium towns. Semi-mechanized aerobic composting is ideally suited to waste volumes in these towns. It demands less in terms of operational and management skills. The product off-take can be good due to close proximity of agricultural areas to almost all such towns in India. The problem of broken glass can be taken care of by suitable local legislation to ensure segregation at source.

Incineration

Incineration is not a total solution for solid wastes. The inert remains still have to be landfilled or used otherwise. This acts as a volume reduction step. In India, it has not found much use as the garbage tends to be low in calorific value and volumes are generally low for a central facility. The technology for incineration is not available indigenously and import options are highly capital intensive. During 1980's an incineration plant was set up at New Delhi at a cost of Rs. 220 million or US$ 6.9 million (May 94). This 300 TPD plant was set up using Danish technology with assistance from Danida. It was also expected to generate power for local grid. The operational experience was not satisfactory. The desired calorific value garbage did not reach the facility as a result of prior segregation due to market mechanisms and scavengers. Despite apparent failure of this attempt, incineration will remain an option for future and experience gained in this venture will be useful. In the meanwhile, incineration on smaller scale with or without energy recovery will continue to be a viable option in a number of location and waste specific cases such as hospital wastes.

Anaerobic Digestion

For high moisture and organic content of Indian wastes, the anaerobic digestion is another suitable option. However, there are no ready technologies available for processing heterogeneous material such as urban solid wastes. The existing methods are suited to homogeneous materials. The costs of cleaning and separating mixed heterogeneous wastes are likely to be high.

Management of solid wastes is of paramount importance in this region because of its increased pollution potential resulting from down stream effects. Rural population relies on surrounding forests for its energy needs. This has been one of the causes of forest depletion. It is difficult to reach conventional urban energy sources such as bottled cooking gas, kerosene etc. to remote rural areas due to difficulties of access in mountain terrain. This makes role of non-conventional energy sources quite important. The rural wastes also assume significance due to their energy potential. The role of simple technologies to ensure efficient utilization of energy from waste will hence be very important in this region. The Bio-gas option will have limited application due to lower prevailing temperatures in the region. It may, however, feasible upto the certain altitudes. Improved wood/bio-mass burning stoves and bio-mass gasification technology can play important role. The urban waste disposal options will be considerably affected by mountain specificities. Land filling may not be possible due to undulating terrain and paucity of flat spaces. Composting may be predominant choice but with due care to intercept run-off from composting areas and its treatment. The seasonal flow of tourists accounts for a good deal of floating population especially during summer. Any planning for solid waste should consider this factor. The large influx of tourists has also resulted in problem of litter at high altitude scenic and tourist spots. This has created peculiar problems of waste retrievals and restoration of those areas.

General Conclusions

1. Bio-gas systems can effectively handle localized and specific wastes and contribute to environment friendly disposal of wastes.
2. The semi-mechanized aerobic composting, however, still remains a strong option for small and medium industrial sectors.
3. Mass incineration will remain a possibility for future.
4. RDF in specific cases is an attractive option provided that sustained indigenous technology development efforts are made.
5. All technologies attempting to process garbage are difficult to master as they are traditionally geared towards handling virgin and homogeneous materials.

Recommendations

The recommendations can be broken up under various categories such as:

- Manpower/Education and Training
- Regulatory and Fiscal
- R & D and Technology Development

Manpower/Education and Training

1. Local governments should work towards infusion of greater management and supervisory skills.
2. Data and knowledge base at the local government level should be strengthened.
3. The problem of over-staffing should be given serious socio-political consideration.

4. Education efforts should focus on women to highlight proper household disposal, segregation and community participation.
5. Steps should be taken to improve health status of scavengers.

Regulatory and Fiscal

1. Local bodies should awaken to the need for suitable legislation as per the prevailing local conditions.
2. Privatization of collection and transportation of urban solid wastes is highly recommended. It will help provide a cap on expenditure, reduce inefficiency and provide better level of service.
3. Private initiatives in waste disposal or utilization should also be encouraged by way of fiscal and other incentives.
4. A nominal garbage tax along the lines of house tax is recommended. It will generate much needed finances and also brings into focus much neglected problem of solid waste.
5. Secondary products should be regulated to protect consumers.

R & D and Technology Development

1. Standard practices for sanitary landfilling under Indian conditions should be developed.
2. Technology and training packages for semi-mechanized aerobic composting at small and medium scale should be developed.
3. R & D efforts should focus on developing improved plants for wastes from vegetable markets, kitchens and restaurants etc.
4. Recycling in the informal sector should be quantified.

REFERENCES

Bhide A.D. *et al.*, Studies on Refuse in Indian Cities, Part II - Variation in Quality and Quality, *Indian Journal of Environmental Health*, Vol. 17, No.3, July 1975, pp. 215-25.

Census of India, 1991.

City Garbage Treatment, A position paper by Business Horizons for Technology Information Forecasting and Assessment Council, Dept. of Science and Technology, New Delhi, India, 1991.

Nagori, G.P. and Rao, C.S., Biogas Manure Plants Based on Agricultural Residues, SPRERI, Vallabh Vidyanagar, India, 1988.

Pickford John, Solid-Waste Problems of Poor People in Third World Cities in 'Practical Waste Management', Ed. J.R. Holmes, John Wiley & Sons, 1983.

Report of Sub Group on Urban Municipal Waste Management, Ministry of Urban Development, Govt. of India, 1990.

Rural Waste Management, A report by National Waste Management Council, Ministry of Environment & Forests, Govt. of India, 1990.

Study on Delivery and Financing of Urban Services by Operation Research Group for Planning Commission of India, New Delhi, 1989.

Swarup, R. *et al.*, Land Holdings' Scenario in the Himalaya, in'Himalaya: Past and Present', Vol. III, Eds. Joshi M.P. *et al.*, Shree Almora Book Depot, Almora, India, 1994.

12

WASTE MINIMISATION IN SMALL SCALE INDUSTRIES

L. Radhakrishnan and *A Saravana Naidu*

Introduction

The quality of environment in India is deteriorating steadily over the past few decades as a result of rapid growth, uncontrolled and lavish consumption, urbanization, industrial expansion and advances in science and technology. Ambient standards of air, water and soil pollution are steadily exceeded with potentially disastrous consequences. India being a developing country, this trends to become more severe everyday with the ever-increasing number of industries.

Small Scale Industries (SSIs) which contribute to a major part of the productivity growth and employment has become collective sources of localised pollution loadings such as organic wastes in water effluent, as well as hazardous wastes like heavy metal sludge, solvents, waste oils, acidic and alkaline wastes, photo wastes etc. Often these wastes are disposed of in an unsafe manner and are extremely difficult to handle it. Thus the waste management problems of SSIs require special attention and special measures and innovative approaches are needed to address their particular problems (Luken *et al.*, 1994). One of the such innovative approachs is waste minimisation. This paper examines some of the techniques of waste minimisation in small scale industries.

Waste Minimisation Concept

Since the beginning of environmental movement in the country, both government regulatory agencies and the industry focused their point where they enter the environment. This concept is appropriately known as 'End of Pipe (EOP) treatment'. While this EOP approach has to some extent been effective for protecting the environment, it has also presented following disadvantages:

- It can result in the transfer of pollutant from one medium to another thereby effecting no net environmental benefits. In some instances this transfer can even increase the risk to human health and the environment.
- It requires huge dead investments and recurring expenses which makes this concept highly unsustainable and at times for most of the SSIs this is just not feasible on account of poor profitability, space constraints etc.

Realising these drawbacks and looking into other pressing problems (the polluted assimilation capacity of the receiving bodies nearing exhaustion) the concerned agencies were forced to look back into the industrial production processes and search for alternative approaches for environmental protection—thus emerged the concept of proactive approach of waste reduction at source as a means for achieving environmental protection. In other words, waste minimisation concept was realised as the need of the day. Waste minimisation can be defined as "A new and creative way of thinking about products and the processes which make them, it is achieved by the continuous application of strategies to minimise the generation of wastes and emissions". For processes, waste minimisation involves conserving raw materials and energy, eliminating the use of toxic substances as much as possible, reducing quantity and toxicity of emissions and wastes before they leave the process etc. For products, it means reducing their environmental impact during its entire lifecycle for raw material extraction till ultimate disposal.

Waste minimisation means economic savings from reduced consumption of raw materials and energy, it means lower pollutant treatment caste, it means better working conditions, it also reflects other benefits such as a better company image. Implementing waste minimisation may not (in fact, will not) solve all environmental problems at a facility, but it does decrease the need for installing and operating end of pipe treatment equipment and reduces the quantity of hazardous waste that needs to be treated and disposed of.

Waste minimisation efforts often reduce worker's exposure to hazardous chemicals, as well as the frequency and severity of accidents and chemical releases. Products that are designed and produced with waste minimisation concepts in mind are often less harmful for consumers to use (Anon, 2004).

Waste Minimisation Techniques

Waste minimisation options or measures could be grouped into two major categories.

1. Source Reduction
2. Recycling and Reuse

Source Reduction

Source reduction is generally perceived as being the highest rung on the waste minimisation leader with the greatest potential for avoiding energy and raw material consumption as well as waste production. With respect to hazardous industrial waste, sources reduction can refer not only to reduction of the volume or weight of waste but also to a reduction of toxicity. The two main types of source reduction are source control and product changes. A more detailed breakdown of these types of source reduction measures can be found in Table 12.1.

As its name implies, product change refers to a change in the nature of the product so that it will produce less waste on release fewer contaminants into the environment when it is used. It does not refer to a direct reduction of the weight of a products packaging (e.g. by substituting plastic packaging for paperboard packaging) but, as illustrated by the example in Table 12.1, can denote a change in the product that will ultimately require less packing.

Table 12.1 : Industrial Source Reduction Measures

Type of Source Reduction Measure	*Description/Example*
Product Chances	Reduce waste/toxicity associated with a product's use
• Product Substitution	Substitute water-based paints for solvent-based paints
• Product Concentration	Concentrate power detergents, thus requiring less packaging
Source Control	Reduce waste/toxicity associated with a product's manufacture
• Input Material Changes	Reduce waste/toxicity of materials used in the production process
Material Purification	Use a higher grade of crude oil during refining, thus reducing the amount of impurities that must be removed
Material Substitution	Substitute water-based cleansers for solvent-based cleansers
• Technology Changes	Reduce waste through process and equipment modifications
Process Changes	Improve the efficiency of chemical reactions
Equipment Changes	Use mechanical scraping systems for cleaning rather than solvents
Process Automation	Automation can optimise product yields by automatically adjusting process parameters
• Good housekeeping Practices	Reduce waste by means of procedural and administrative measures
Management and Personal Practices	Offer employee education programmes, bonuses and awards to encourage employees to reduce waste
Waste Stream Segregation	Facilitate recycling by preventing mixing of different waste types, particularly hazardous and non-hazardous wastes
Inventory Control	Use input materials before expiry dates
Loss Prevention	Check for spills and fix leaks from equipment
Cost Accounting	Allocate waste treatment and disposal costs directly to the departments or groups that generate the waste

Source: United States Environmental Protection Agency (1998), Ghassemi (1989).

Most source reduction measures fall into the category of source control. These are measured that reduce the waste and/or toxicity associated with a product's manufacture. Input material changes are source control measures which include the purification of raw material inputs and the substitution of less toxic materials.

Technology changes for source reduction include changes to the design of the production process, the purchase of new equipment, changes in operating conditions and changes in process controls.

Good housekeeping or good operating practices refer to changes in the procedural, organisational or institutional aspects of waste management. Inventory control is a housekeeping measure that controls the quantity and type of materials found in a plants inventory. Inventory control measures seek to reduce the amount of waste resulting from materials that are out of date, off-specification, contaminated, damaged, or excess to the plant's needs. Like other housekeeping measures, inventory controls are relatively inexpensive and easy to implement (Maclaren, 2000).

Recycling and Reuse

Waste materials generated during the production process can be reused or recycled both on-site (either in the plant or on the plant property) and off-site. "Reuse" means reusing a waste material directly, either for its original purpose or in a new role, without any major modification to the material before it is reused.

Waste materials that are "recycled" require some form of significant physical, chemical or biological processing. For example, used paper must be denied, pulped and remanufactured before it can be used again.

An important aspect of reuse and recycling in developing countries that differs substantially from practices in developed countries is that developed economics tend to favour more extensive recycling and reuse of materials. On the demand side, the value of raw material inputs tends to be higher and their availability.

Scarcer, thus making recycled or reused materials more attractive. On the supply side, the labour costs associated with separating mixed wastes are also very low. Industrial waste material sorting for the purposes of reuse and recycling is an important income generating source that can benefit a wide range of actors, including factory owners, factory employees, scavengers, junk buyers, and municipal waste collection employees (*Ibid.* : 4-5).

Small Scale Industries and Waste Minimisation

Definition of Small Scale Industries

1950	:	First definition of Small-Scale Industries (SSIs) : Investment limit fixed at Rs.5 lakh in fixed assets and employing less than 50/100 persons with or without power respectively.
1960	:	Definition of SSI was changed. Employment clause dropped.
1985	:	Investment limit was increased to Rs.35 lakh in plant and machinery.
1991	:	Investment limit increased to Rs.60 lakh.
1997	:	Investment limit increased to Rs.3 crore.
1999	:	Investment limit was brought down to Rs.1 crore.
October 2001	:	The investment limit in plant and machinery for specified hosiery and hand tool items was raised to Rs.5 crore.

There are about 3.4 million SSIs operating in India employing more than 19.2 million people in which 86 per cent of employment is in the manufacturing sector alone. It ranks next to agriculture in terms of employment generation in the country. The SSIs are the backbone of India which accounts for 95 per cent of industrial units and 40 per cent of industrial output worth Rs.700,000 crore and contributing about 7 per cent to the GDP. The contribution of SSIs to exports is between 45-50 per cent. The concentration of SSIs in India: Uttar Pradesh (12.51%), Tamil Nadu (11.34%), Madhya Pradesh (9.71%), Maharashtra (7.83%), Karnataka (7.30%) and Punjab (6.38 %). The state- wise employment is that Tamil Nadu (20.02 %), Maharashtra (10.28%), Uttar Pradesh

Tamil Nadu (20.02 %), Maharashtra (10.28%), Uttar Pradesh (9.73%), Karnataka (7.39%), Delhi (7.38%) and Gujarat (6.46%): The SSIs are producing output of more than 7,500 products.

Above all, these SSIs contribute about 65 per cent of industrial pollution, generating about 3881 million litres of waste water per day. Recalcitrant organics and inorganic substances compound the problems of treatment and safe disposal. Unplanned industrial agglomerations and apparent lack of adequate resources to enforce and implement pollution control problems are implicated as some, of the major reasons for the prevailing chaos (Yadav, 2002).

Table 12.2 : Poisoned Waters [About 40 per cent (3,881 million litres per day) of waste-water generation comes from small-scale industries]

Industry	*Cluster Regions*	*Waste-water generated (million litres per day)*
Engineering	Hand Tools Jalandhar and Nagaur, Locks Aligarh, Scientific Instruments Ambala, Ajmer, bicycle and parts, Ludhiana Brass parts, Jamnagar diesel, engines and parts, Kothapur, Agra, Rajkot, Coimbatore Ghaziabad.	2,125
Paper and board mills	Spread across the country	1,087
Textile industries	Woollen and Cotton Hosiery Tripur, Ludhiana, Kolkata, Delhi	450
Organic chemicals	Gujarat	60
Tanneries	Leather and Leather Products Chennai, Ambur and Vaniampadi (Tamil Nadu), Kolkata (West Bengal) and Agra and Kanpur (Uttar Pradesh)	50
Pharmaceuticals	Gujarat and Maharashtra	40
Dye and dye intermediates	Gujarat	32
Soaps and detergents	Gujarat	10
Paints and varnishes		10
Petrochemicals		10
Edible oil and vanaspaty		7

Source: Anon, 2001, Pollution Control in Small-scale industries—Status and Needs, Central Pollution Control Board, January 2001.

According to a status report on SSIs by the Central Pollution Control Board: "The level of pollution caused by the SSI sector in India per unit of output is higher than their counterparts in developed countries". The reasons are many:

- Continued usage of outdated and inefficient technologies that generate large amounts of wastes;
- Large and unplanned industrial conglomerations;
- Lack of resources for enforcement and implementation of pollution control programmes;
- Lack of public/market pressure for improving environmental performance; and
- Lack of proper siting thereby posing greater environmental risks.

The Union Government's attempts, through few and feeble to clean the sector has failed. In the late 1990s, the Union Ministry of Environment and Forests (MEF) in collaboration with World Bank launched a project for waste minimisatoin among SSIs. Forty-five waste minimisation circles were set up in various sectors in different parts of the country. But mostly these projects largely remain restricted to demonstrations and pilot projects and have achieved partial success. Out of the 300 waste minimisation options identified, only 120 have been implemented. A recent project in the town of Tirupur, Tamil Nadu, failed to take off, as the industrialists showed little interest. The governments focus then shifted to Common Effluent Treatment Plants (CETPs).

Examples of Indian Industry Becoming Clean

Some experiments have proved that process modification can lead to increased efficiency and reduce pollution. One such experiment is being carried out by the New Delhi based National Cleaner Production Centre (NCPC) in the hosiery cluster at Ludhiana, Punjab. Earlier the unit generated a large amount of waste due to wet processing and also due to several factors such as obsolete and

Table 12.3 : Common Effluent Treatment Plants (CETP) in India

State	*Ongoing CETP project*	*Operational CETPs*
Andhra Pradesh	2	3
Gujarat	4	10
Haryana	2	1
Himachal Pradesh	4	-
Karnataka	-	3
Maharashtra	10	6 (plus 2 contracted but not operating due to lack of effluents)
Madhya Pradesh	3	-
Delhi	15	-
Punjab	4	1
Rajasthan	2	3
Tamil Nadu	34	21
Uttar Pradesh	-	3
West Bengal	3	-

Source : Anon 2002, Parivesh Special Issue on Common Effluent Treatment Plants, New Letter of Central Pollution Control Board, New Delhi.

in efficient technology, poor operation and maintenance practices, lack of production planning, and non-availability of information.

The project used various cleaner production techniques that included better housekeeping, input material change, recycle/reuse, technology change, better process control, equipment modification and useful by-product formulation. For example, there is a tendency to use 5-10 per cent excess chemicals as a safety factor. But only a very small percentage of chemicals, added during processing, are consumed and the rest finds way into the waste water.

The results have been amazing. The project has demonstrated that wastewater can be reduced by 10-80 per cent, air pollution by 6-47 per cent and hazardous and solid waste by 3-12 per cent. Moreover, the profit enhancement potential for hosiery processing units is in the range of Rs.2,000 to Rs.5,000 per tonne of fabric processed (*Ibid.* 28-34).

The waste minimisation circle initiations have been particularly successful in mobilizing participation of industry in reducing wastes and improving energy efficiency in particular. These include initiatives in the rice mills (Kumar, 2000), the distillery industry (Patil, 2000) and stone crushing operations (Ravishankar, 2001). Reduction in water consumption, chemicals and dyestuff, optimization of fuel efficiency and electrical energy conservation have lead to significant saving to industry in several of these cases.

Water stream analysis has helped reduce energy losses and water consumption in safe and dairy industries (Nair, 2001). Nearly 17 waste minimisation options were identified in electroplating operations that helped guide action in several firms (Mahulkar, 2002). Saxena (2002) presents the case for analysing green productivity interventions in edible oil industry that can help reduce oil, hexane and energy losses in particular in edible oil industry.

Recently waste minimisation initiations in the fertilizer manufacturing industry has also been reported, that could help save on water, magnesium oxide and sulphuric acid in particular (Nair, 2002). Significant improvement in quality has also been reported in these cases (Gopichandran *et al.*, 2003). These are some of the success projects using waste minimisation techniques.

Conclusion

Though small-scale industries in developing country like India accounts for a large share of employment and production but at the same time in terms of pollution load, many of the, SSIs are major hazardous waste producers and certainly contributing to environmental pollution and serious occupational exposure. Among these are tanneries, textile dyeing plants, dyestuff producers, electroplating shops, pesticide formulation, paint shops and printing establishments. In this scenario waste minimisation have emerged as a sensible approach for sustainable industrial development. And over the years waste minimisation techniques has proven to be an effective way to control pollution from small-scale industries.

REFERENCES

Anon (2004), "Waste Minimisation Benefits, Legal Aspects, policy instruments for pollution abatement", *Waste Minimization Cricle*, National Productivity Council of India.

Gopichandran, R. and Praveen Prakash (June 2003), "Environmental efficiency means greater economic benefits and improved competitiveness: Lcssons for Indian Chemical Industries in Particular", *Environmental Pollution Control Journal*, 6(4): 46-47.

Kumar, I.V.R. (Aug. 2000), "Waste Minimisation in Rice Mills", *Waste Minimization Circle News*: 3-9.

Luken A. Ralph & Ritukumar (1994), "Waste Minimisation Programme for Small-scale Industries", *Productivity*, 35(1): 44.

Maclaren W. Virginia (2000), "Appropriate Industrial Waste Management Strategies for Developing Countries", *Waste Economy,* University of Toronto: 3-5.

Mahulkar, R.C. (June 2002), "WMC in electroplating industry", *Waste Minimisation Circle News*: 3-5.

Nair, N.G. (Dec. 2000), "WMC case studies from sage and dairy industries", *Waste Minimisation Circle News*: 6-12.

Patil, S.M.S.M. (Aug. 2000), "Waste Minimisation in distillery industry through conversion of batch distillery to continuous process", *Waste Minimisation Circle News*: 9-11.

Ravinshanker, D. (Sep. 2001), "Cleaner Production in Stone Crushing Unit— A Case", *Waste Minimisation Circle News*: 4-8.

Saxena (Sep. 2002), "Green Productivity Demonstration in Edible Oil Industry —A case", *Waste Minimisation Circle News*: 4-7.

Yadav, K.P.S. (2002), "The small big polluter", *Down to Earth*, 11(10); 26-34.

13

BIO-WASTE FOR SELECTED TEXTILES USE

I. Jayalakshmi

Environmental awareness and consciousness is increasing all over the globe, posing a new challenge to the textile industry. During the last two decades, there has been a growing awareness of the environmental damage caused by the indiscriminate use of toxic chemicals, in the textile wet processing units. Some of which are very toxic and carcinogenic in nature. The effluent discharged from these wet processing units cause skin and lung cancer, allergy, anemia, renal damages besides hearing and visual impairment.

An awareness of the pollution problem has forced the textile industry to think of eco-friendly means of textile production and wet processing. The concern of environment and hunger for eco-friendly products and process have compelled the scientists/technologists to think of using natural means for textile processing. It is in this context that natural dyes are gaining popularity today.

Dyeing textiles with natural dyes is an age old practice and a rich tradition of India. Natural dyes are an ideal choice because they have bio-degradability. It constituents are anti-allergens and safe for body contact. Also, naturally dyed materials are non-toxic and non-hazardous to health.

The acute ecological crisis has caused the environmentalists to raise the call : "*Go back to Nature, if the World is to be saved from poisoning*". Interest in natural product is growing throughout the world and people are becoming aware of the need for eco-friendly materials to come up and dominate the scene.

According to Chavan (1995), the use of natural dyes on textiles has been just one of the consequences of increased environmental awareness. Naturally dyed materials have good resistance to moth invasion. Some of its constituents are anti-allergens and hence prove to be safe for body contact. Naturally dyed materials are non-toxic and therefore not hazardous to health.

Hummel (1993), classified the colouring matter as monogenetic and polygenetic. Monogenetic dyes produce only one colour irrespective of the mordant present on the fibre or applied along with the dye. Polygenetic colours produce different colours according to the mordant employed.

Hence, an attempt has been made to extract dye from a selected vegetable source and apply the same on silk fabric.

> The main *objective* of the study is to observe the Response of Silk Material for Natural Dyes, Analyse the Fastness of Dye and Performance of the Dyed Material.

It is hoped that this study would add to the eco-friendly character of silk fabrics.

The *Methodology* adopted for the study consisted of the following steps:

1. Selection of Fabric
2. Selection of Dye Source and Mordant
3. Optimising Dyeing Procedure
4. Dyeing of the Samples
5. Evaluation of the Dyed Samples

1. Selection of Fabric

Silk being an eco-friendly material was selected for the study. Silk is known for its elegance, soft feel and luster. Moreover, silk is a traditional fabric, loved by almost all women.

Preparation of Fabric

Silk purchased for the shop was stiff owing to the presence of gummy substance called sericin. It needs to be removed before dyeing as it hinders the dye absorption. This process is known as degumming. Silk fabric was soaked in soap nut water and boiled for half-an-hour to remove the sericin (gum/starch substance).

2. Selection of Dye Source and Mordant

Natural dyes, namely Pomogranate rind (Yellow/red), French marigold flower petals, Onion skin, Teak leaves, Grape juice extract were tried in the pilot study. Based on the results, *Pomogranate rind and French marigold flower petals* were selected.

Pomogranate with yellow and red rind, were dried in the shade separately and powdered.

French Marigold flower petals were also dried in the shade and powdered.

Natural dyes need the use of mordants for colour fixation. Mordants can be natural or synthetic. *Myrobalan, an eco-friendly natural mordant* was taken for the study.

Gulrajani *et al.* (1993) suggest three mordanting techniques, namely—

1. Pre-Mordanting
2. Simultaneous Mordanting
3. Post-Mordanting

Pre-Mordanting is a technique in which the fabric is first soaked in mordant first and then boiled in the dyestuff.

Simultaneous Mordanting is a technique in which the mordant as well as the dye is mixed together and the fabric is soaked and boiled, in this mixture.

Post Mordanting is a technique in which the fabric is first boiled in the dyestuff and then soaked in the mordant.

3. Optimising Dyeing Procedure

The pilot study proved that simultaneous mordant was suitable for promogranate dye while pre-mordanting was suitable for French marigold.

The *dye solution* was prepared as follows :

Requirement	*Pomogranate Rind (Yellow/Red)*	*French Marigold*
Fabric Weight	30g	30g
Dye Powder	300g	300g
Mordant	Myrobalan	Myrobalan
M: L	1:20	1:20
Water	2,500 cc	3,500 cc
Temperature	30 min	60 min
Time	90° C	90° C

Preparation of Dye Solution

The weighted quantity of the respective dye powder was dissolved in 3,500 of water and boiled for 60 min., at 90°C. Later the dye solution was decanted carefully for further use.

4. Dyeing of the Samples

Pilot study revealed that *simultaneous mordanting* was suitable for pomogranate, while *pre-mordanting* was ideal for *French marigold.*

The red and yellow pomogranate rind and Myrobalan powder were dissolved in water and brought to boil separately. The degummed silk material was entered into the respective dye solution and boiled for 60 min. Later, each of the samples was taken out and rinsed thoroughly in three changes of soft water and dried in the shade.

As regards for French marigold, the degummed silk sample was first boiled for 60 min in myrobalan solution and then boiled in French marigold dye solution for 60 min. Later, the sample was taken out and rinsed thoroughly in three changes of soft water and dried in the shade.

5. Evaluation of Dyed Samples

The dyed samples were evaluated both *visually* and also through *laboratory tests* for assessing *strength, thickness, drape and colourfastness*. The results thus obtained were analysed statistically.

Visual Evaluation

The silk material which was dyed by pomogranate rind and French marigold flower petals proved to be *good in general appearance, softness and luster.*

Objective Evaluation

The findings of the study are as follows :

> Samples dyed with *pomogranate rind* had *good colour fastness to sunlight, washing, pressing and crocking.* While the samples dyed with *French marigold* had *excellent fastness to light and dry pressing*. But the *fastness to washing, wet pressing and crocking was only moderate.*

Strength Text

The results of the strength test are discussed in Table13.1.

Table 13.1 : Strength Test

Samples	*Warp*		*Weft*	
	Mean Strength in Kg	*% Loss or Gain over Original*	*Mean Strength in Kg*	*% Loss or Gain over Original*
Original	27		31	
Red Pomogranate	41	151.8	34	109.6
Yellow Pomogranate	39	144.4	32	103.2
French Marigold	27	100	34	109.6

In *warp*, the silk samples which were dyed by *Red Pomogranate rank first* in strength, followed by Yellow Pomogranate and French marigold. Whereas, in *weft Red Pomogranate and French marigold* has good strength than *Yellow Pomogranate*.

Thickness Test

The results of the thickness test are discussed in Table 13.2.

Table 13.2 : Thickness Text

Samples	*Mean Thickness in mm*	*% Loss or Gain over Original*
Original	15	
Red Pomogranate	23	153.3
Yellow Pomogranate	18	120
French Marigold	17	113.3

The silk material dyed by *Red Pomogranate* shows better thickness than French Marigold.

Drape Test

The results of the drape test are discussed in Table 13.3.

Table 13.3 : Drape Test

Samples	*Mean Drape Coefficient %*	*% Loss or Gain over Original*
Original	20.6	
Red Pomogranate	18	87.3
Yellow Pomogranate	18	87.3
French Marigold	18	87.3

The silk samples dyed by Red Pomogranate, Yellow Pomogranate and French marigold have a Marginal drape co-efficient.

The study reveals that bio-waste, namely Pomogranate rind and dry French Marigold flower petals can be very good source of natural dyes. Even a costly material like silk, can be dyed with these bio-wastes. Definitely wealth can be obtained from waste.

REFERENCES

Chavan, R.B. (1995), "Revival of Natural Dyes, a word of caution to environmentalist, Clothes line," Vol. 8, No. 4, pp.36-100 Vol.42, No.4, pp. 27 - 30.

Hummel (1993), "Compendium of Inter-Religious Workshop on Natural Dyes," NHDC Ltd., Lucknow, p.89.

Gulrajani, M.L. (1993), "Introduction to Natural Dyes and Mordants", Compendium of Inter-Religious Workshop on Natural Dyes, NHDC Ltd., Lucknow, p. 82.

14

STRATEGIES FOR WASTE MINIMISATION IN SMALL SCALE INDUSTRIES

S. Syed Ahamed and *A. Baradeswaran*

Introduction

A business is composed of three distinct and different sources such as (a) a set of non-living things, viz. plant and machinery, raw material, spares, parts, building etc., (b) a group of living organ especially human being, viz. workers, supervisor, managers executive, directors, managers, executive, directors, supervisors, chairman etc., (c) Environment which means our surroundings in which we took birth, survived, grew, now share, and will perish. God created the environment, but environment is the creator of a business opportunities. So the business has got the obligation to protect the environment.

Cost Imposed by Business on Environment

- Soil erosion and ground water loss
- Destruction of the rain forest
- Acid rain
- High level of air pollution
- Depletion of natural resources
- Spoiling water of sacred rivers
- Garbage and solid wastage
- Loss of ozone layer

Benefits Rendered by a Business Corporation to Environment

- Green lawns and social forestry

- River purification projects
- Greening barren land

Benefit Received from the Environment

- Sunlight—vital source of renewable energy yet free of cost
- Water—versatile renewable energy yet has no or little economic value
- Oxygen and carbon dioxide—live fuels
- Provider of important of raw material
- Human resource supply
- Rain, snowfall, icebergs
- Sand and clay

Cost Imposed by Environment

- Flood and droughts
- Cyclone, storms, typhoons
- Earthquakes and volcanoes

Role of Small-scale Industries

Small-scale industries account for 40 per cent of total industrial production, 35 per cent in total export, and employing 3.2 million industrial units they are involved in food reprocessing, hosiery and garment, and leather and leather products. They contribute 40 per cent of the total industrial waste in India. There are many ways to reduce industrial waste in following ways.

Waste Minimisation Techniques

Waste minimization includes any source reduction and/or recycling activity undertaken by a waste generator. These activities result in a reduction of waste produced and/or, a reduction in the toxicity of the waste. Some examples of waste minimisation techniques are listed below.

Source Reduction Techniques

- Change the composition of the product to reduce the amount of waste resulting from the product's use.
- Reduce or eliminate hazardous materials that enter the production process.
- Use technology to make changes to the production process; equipment, layout or piping; or operating conditions.

Good operating practices such as waste minimisation programmes, management and personnel practices, loss prevention, and waste segregation help to reduce waste at their source.

Recycling Techniques

- Return waste material to original process.
- Use the waste material as a raw material substitute for another process.
- Process waste material for resource recovery.
- Process waste material as a by-product.

Waste Auditing

A waste audit

- Defines sources, quantities and types of wastes generated;
- Identifies where, when, how and why these wastes are produced;
- Identifies areas of wastage and waste problems; and
- Establishes targets and priorities for waste reduction.

The waste audit can be used to

- Ensure external regulatory compliance;
- Develop base-line data; and
- Evaluate alternatives to minimize wastage of resources.

Cleaner Production

The United Nations Environment Programme (UNEP) coined the term Cleaner Production when it launched the Cleaner Production Programme in 1989.

What is Cleaner Production?

Cleaner Production is the continuous application of an integrated preventive environmental strategy applied to processes, products, and services to increase overall efficiency and reduce risks to humans and the environment (UNEP, 1989).

- For production processes—the strategy includes conserving raw materials and energy, eliminating, toxic raw materials, and reducing the quantity and toxicity of all emissions and wastes.
- For products—the strategy focuses on reducing negative impacts along the lifecycle of a product, from raw materials extraction to its ultimate disposal.
- For services—the strategy involves incorporating environmental concerns into designing and delivering services.

Cleaner production requires changing attitudes, responsible environmental management and evaluating technology options.

Benefits of Cleaner Production

- Economic benefits include increased profits; improved marketability; lower expenditure on materials, resources, waste disposal and transport; additional sources of income: increased quality and greater productivity; better customer service; and increased goodwill from being seen as a responsible member of the community.
- Environmental benefits include less pollution; reduced production of waste; more efficient use of resources; less

waste going to landfill; and less effluent to sewage or discharges to water bodies.

Key Factors in the Success of a Cleaner Production Programme

1. Commitment of Management and Employee Involvement— The involvement of all levels of an organization is vital to the success of a programme.
2. Openness to Change—Overcoming perceived barriers such as "we've tried that before and it did not work" or "this is the way we have always done it", through group discussions and "brain storming sessions", when all options and their feasibility are openly discussed.
3. Availability of Resources—It is important to establish at the start of a programme, the economic constraints and availability of human and other resources.
4. Planned Approach—The complexity and detail of the planning process will depend on the project(s) and the availability of an organization's resources.

Potential Barriers—Before implementing a programme, it is worth looking at some difficulties that may be faced, as usually the organization has some control over these barriers.

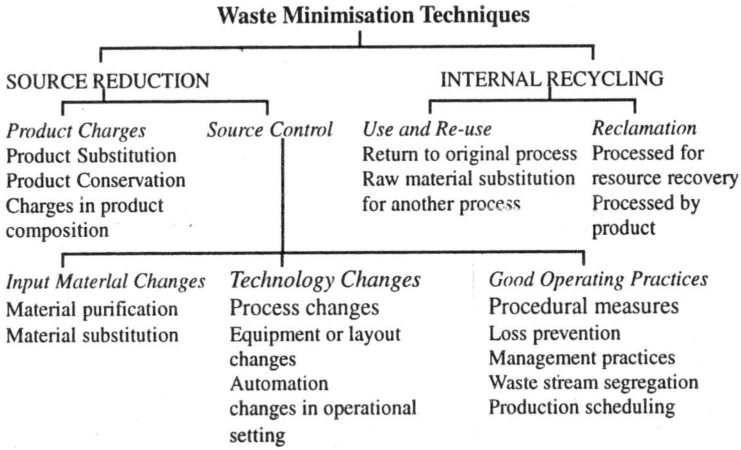

(Adapted from US EPA's Waste Minimization Opportunity Assessment Manual)

A Case Study of Cleaner Production in Brewery on the Island of Mauritius

The brewery was established in 1963. It employs about 500 persons and the annual production is now just over 35 million litres of beer. Most of the production is meant for the local market but the export share is increasing. The company is expecting to double its annual production in the coming years, and to succeed in achieving this target the management is investing a lot in modern production techniques and training for its personnel. The factory discharges its effluents directly into the sewerage network without any pre-treatment. The solid wastes generated are collected by lorries and disposed in a landfill. Broken glasses are recycled and spent grains, dried yeast, and trub are sold to other companies to be used as animal fodder.

Emissions and Material Balance

The water consumption is of the order of 1200 m^3/day. The main outputs from the manufacturing process are mainly—

Beer	120 m^3
Wastewater	900m^3
Solid wastes	10,000 kg/day
Broken Glass	2,000 kg/day
Spent grains and trub	32,400 kg/day

The total wastewater is about 900m^3/day and the pollution load is about 5,500 kgCOD (Chemical Oxygen Demand) /day and 490 kg TSS(Total Suspended Solids)/day. The brewing and processing operations generate the higher-strength effluents. Wastewater streams from the brew house and the fermentation areas were found to contain the highest organic and suspended solids loads. The overall pollution load was 5,500 kgCOD/day and 4,920 kg SS/day. Local farmers for land preparation collect the 4,000 kg of ash from the boiler blow-down. The spent grains and trub are sold to local companies as animal feed material whereas the broken glass is recycled in a sister company.

All the CO_2 produced during fermentation is recovered, stored, and reused. Some is sold to external companies when required. Stack emissions from the boiler house contain particulate matter and sulphur oxides. The boilers use heavy fuel oil with a sulphur content of about 3 per cent. The stack heights are low and the particulate emissions are above the permissible maximum limits. Residents close to the brewery complain mainly about noise and particulate matter.

Cleaner Production Opportunities

The brewery consumes about 6.0×10^6 MJ of energy monthly (100000 litres of heavy fuel oil and 140 tonnes of Coal). The brewery is already adopting a few clean technologies such as:

- Recovery of CO_2
- Partial recovery and drying of yeast
- Counter-current use of water in washings
- Heat recovery and hot water storage tank
- Reuse of caustic soda

Analysis of various CP options showed that it was technically and economically feasible to reduce the water to beer ratio by one third. Furthermore, by re-circulating the trub, spent grains and residual beer, into the process, about 8 per cent of additional beer can be produced, leading to a reduction in the pollution load.

Gujarat Cleaner Production Centre

The concept of waste minimisation one step ahead of waste management gave birth to Gujarat Cleaner Production Centre (GCPC). Cleaner Production means pollution prevention or waste minimisation with increase in production with minimum wastes and emission for more competitive and profitable industry along with environment protection. In these regards and due to rapid industrialization in the state, GCPC has been established in the corporation in August 1998 with the technical support of UNIDO and National Cleaner Production Centre (NCPC), Delhi and with financial support of GIDC. Systematic application of Cleaner Production tools like Good House-keeping, Recycling and Process Modification has benefited industry on various fronts like,

- Improvisation in efficiency
- Lower Costs
- Conservation of Raw Material and Energy
- Better Access to Finances
- Improvisation in Market Requirements
- Improvisation of Environment
- Better Compliance with Environment Regulations
- Improvisation of Company and Public Image

The Centre has been established to create awareness for C.P. among the Industries in Gujarat by organizing Orientation Programs and Workshops within the State for educating the industries towards Cleaner Production. It is aimed to make the industries realize the importance of Pollution Prevention as compared to Pollution Control, which would help them in increasing their profits as well as protect the environment.

Protection of Natural Capital as a Business Imperative in Australia

Visy Industries—Paper and Packaging

Visy Industries has established a $400 million dollar plantation pine pulp and paper mill at Tumut in New South Wales. The mill will produce 240,000 tonnes of unbleached Kraft pulp and packaging paper for both domestic and export purposes. Through the integration of environmental considerations into the design and operation of the mill arrange of benefits have been delivered:

- Native forests will be preserved through the use of softwood plantations;

- Intensive use of commercial and domestic waste paper further serves to protect native forests;
- The mill will be powered by hydro-electricity and on-site generated biomass fuel to minimize GHG emissions;
- Water usage will be the lowest per tonne of pulp produced of any mill in the world.

Processes have been specifically designed to minimise requirements for chemical use. The focus has been on achieving both eco-efficiencies and resource input substitution to produce a sustainable paper and package recycling business operation. In doing so a number of ongoing environmental benefits will be realized.

Source: Visy Industries website <URL http://www.visy.com.au> (last viewed at September 24, 2001).

Implementation of BP Global Choice TM Programme

It has been estimated that the average car on Australia's roads travels about 16,000 kilometres each year, requiring 1,920 litres of fuel and resulting in the emission of approximately 4.33 tonnes of CO_2 into the atmosphere. With about 10 million vehicles, currently in use this equates to approximately 43 million tonnes of CO_2 emissions each year. As part of an integrated strategy to reduce the environmental impact of its operations, BP Australia has introduced the BP Global Choice TM programme. Each time a customer fills up their vehicle with BP Ultimate fuel, BP Australia will offset the equivalent level of greenhouse gases emitted by the vehicle by investing a percentage of the purchase price of the fuel into a range of independently audited greenhouse gas reduction projects.

Examples of projects funded by the BP Global Choice TM programme include:

- Renewable energy (clean energy from solar, wind, tidal and hydro);
- Greenhouse gas captures initiatives (i.e. turning methane gas from landfills into electricity);

- Planting new forests to absorb CO_2 from the atmosphere; and
- Implementing innovative technologies to reduce the levels of pollution caused by existing manufacturing and other processes.

It is understood by BP that the BP Global Choice Programme is the first of its kind in the world linked to a petroleum product, and subject to continued customer support, the company is committed to expanding access to this innovative product within the Australian market place.

Source: BP Australia 2001, Website<URL: http://www.bp.com.au/globalchoice> (website accessed 25/09/2006)

15

MARKETING STRATEGIES AND ISSUES IN MICRO ENTERPRISES OF SELF HELP GROUPS

M. Soundarapandian and *G. Saravanan*

Self help group is a voluntary group formed to attain some common goals. Most of its members have similar social identity, heritage, caste or traditional occupations and come together for a common cause and manage resources for the benefits of the group members. These groups as they provide monetary and moral support to individual members in times of difficulties.

The self help group can be defined as "a homogenous group of rural poor voluntarily formed to save whatever amount they can conveniently save out of their earnings and mutually agree to contribute to a common fund from which to lend to members for productive and emergency credit needs."

The self help groups aim at providing awareness among the poor about the ongoing development programmes. The poor should know how best to use existing governmental programmes and also the legal provisions meant for the disadvantaged sections of the rural communities.

Under Swaranjayanti Gram Swarozgar Yojana (SGSY), the beneficiaries are as Swarozgar. The Swarozgaris can be individual or groups. SGSY lays emphasis on the group approach. Under the scheme, the rural poor are organized into self help groups in either the list of BPL household identified through BPL census, duly approved by the Grama Sabha will form the basis for identification of families for assistance under SGSY. The self help groups also be drawn from the BPL list approved by Grama Sabha.

Small enterprises includes running poultry farms, rearing ducks, goat rearing, bullock rearing, pettyshop, ration shop, making vermibed, making coir, rope making etc. Rural women are trained on domestication of small ruminants before they are introduced to the same.

Rural women are now undertaking clay modelling, preparing bamboo baskets, plate making with shal leaves, bidi making, belmala and Kurchimala preparation and batic and fabric print on clothes, wool knitting, etc. as a part of their income generating activities. The most interesting fact that all the women are imparted training in the concerned field before undertaking the enterprise.

In all the enterprises women are making sincere efforts to achieve success in near future. Those who have undertaken the small enterprises for commercial purpose are being benefited in different ways, either monetary or in kind. Also a greater number of women are now taking part in different types of rural development programmes. It is because of that they are conscious and aware about various issues relating to rural development in general and women's development in particular.

The benefits of rearing small ruminants are:

Domestic consumption	—	Monetary benefit
Resource mobilization	—	Social recognition
Manpower utilization	—	Building marketing skill
Experiential learning	—	Bargaining capacity
Confidence building	—	Dealing with customer

An attempt has been made to analyse the Marketing Management of Micro Enterprises of SHGs in Dindigul District of Tamil Nadu.

Objective of the Study

The specific objectives of the study are,

1. To study the functioning of micro enterprises of SHGs in Dindigul District;
2. To analyse the performance in marketing management of micro enterprise of SHGs in the District;

3. To examine the problems of the micro enterprise of SHGs in study District;
4. To suggest the concrete measures for effective marketing management of micro enterprises in Tamil Nadu State.

Methodology—Design of the Study

The present study is based on primary as well as secondary data. The district level secondary data was collected from the District Rural Development Agency. Lead Bank, and Mahalir Thittam at Dindigul. The data for block level performance of micro enterprise of SHGs were collected from the office records of Block Development Office and from the records of the panchayat offices. The publications of Government of India, Government of Tamil Nadu and various evaluation reports were referred for the present study. Various journals, newspapers, magazines and books related to marketing management, micro enterprises and SHGs were also gone through for the present study.

Dindigul, Pillaiyarnatham, Samiyarpatti, Gandhigram, Panjampatti, Settiyapatti and Velankannipuram of Dindigul District was purposively selected for the study in order to analyse the performance of marketing management in micro-enterprise of SHGs.

At the respondent level after selecting the sample village panchayats, a list of micro-enterprise of the SGSY programmes during the period from 1999-2000 to 2003-2004 were prepared from the records of Block Development Office. 20 sample enterprises were randomly selected for the interview purpose. A total of 100 SH group members were administered with a protested questionnaire.

Micro-Enterprises of SHGs

In micro-enterprise sub-sector, the entry as well as exit costs are lower. However these costs are formidable for the resource poor wage earners who set their aim to make it into self-employment.

That also explains their high risk perception. But, then as the micro-enterprises grow in size they trend to explore and even move into areas of higher value addition and better earnings.

Two market strategies, viz. market penetration and market creation could be followed for identifying the self-employment potential. The penetrated market is the set of consumers who are buying the company's products, of these market-penetration merits the first mention. Under this strategy:

- The new entrepreneur is trained to enter the market essentially on non-price considerations usually the focus is on physical placement in the market.
- Wage-earners apprentice, who comprehend the market well but are not able to venture into self-employment in the industry due to capital constraints hold a better change of success. The threshold level capital assistance cold help them to find entry into the market.
- New compositions of market structure could emerge and more micro-operators could join the existing ones.
- More self-employment avenues should be created.
- Price-competition could increase gradually.
- Since the existing market is redistributed only ownership transfer takes place.
- At the macro level the size of the market remain unchanged.

A creative market discovers and produces solutions. Customers did not ask for but to which they enthusiastically respond. Market creation is second strategy worth exploring of these two the latter is challenging as no guide models are available. It has higher incidence of risk too. The application of this strategy calls for comprehensive assessment of market. In appraising the projects conceived on this market strategy, the significance of relatively longer gestation periods and difficulties in quitting the market. In case of failures have to be appreciated. In this strategy:

- Returns on investment even for the most viable enterprises, have to be considered with long-term perspective.

- Usually, first generation entrepreneurs should avoid such strategies.
- If the product/service clicks the entrepreneurs to have the advantage of being pioneers in the trade. Competition is usually non-existing until the second generation operators respond to the opportunities.
- Enterprise under this strategy help in creating new jobs and new activities in the area. They help in giving a fillip to the local economy.
- At the macro-level, they contribute to the over all production and increase in employment.

Marketing mix relate to four basic factors, such as (i) Extent of satisifaction of the people (customer/consumer), (ii) Value of the product converted into price to be paid by them, (iii) Cost of production followed by supply/distribution involved and finally, (iv) The quality of utility of the same claiming choice of preference before the target market, aimed at, to capture customer/consumer base. Thus, an enterprise depend upon place, product, price and promotion.

The four P's (place, product, price, promotion) from the seller point of view constitute the effective marketing mix. From the customers point of these four C's indicate (EPDP)—

- Customer's Comfort (Effectiveness),
- Cost to the Consumers (Price),
- Convenience to the Customers (Distribution),
- Communication to the Customers (Publicity).

Table 15.1 shows the marketing mix tcchniqucs followed by the sample micro-enterprises in the Dindigul District. While selecting the place mix, out of 20 micro-enterprises, 15 enterprises are more concentrated in location and coverage of dealers. With regard to transportation, the enterprise concentrate more or less equally. While selecting the product mix, nearly 16 enterprises, are less concentrated on variety, brand name, design, size and

Table 15.1 : Marketing-mix in Micro-Enterprises

Sl. No.	Factors	Concentrated	No. of Micro-Enterprises More Concentrated	Less Concentrated	Total
Place					
1.	Location	5	15	–	20
2.	Coverage of dealers	4	12	4	20
3.	Inventory level	2	6	12	20
4.	Transportation	6	8	6	20
Product					
5.	Variety	2	2	16	20
6.	Quality	7	8	5	20
7.	Design	3	3	14	20
8.	Brand Name	2	2	16	20
9.	Size	3	3	14	20
10.	Packaging	5	5	10	20
11.	Guaranty	1	2	17	20
Price					
12.	Quoted Price	14	5	1	20
13.	Discounts Allowed	2	9	9	20
14.	Trade is Allowance	3	6	11	20
15.	Credit Period	10	8	2	20
16.	Payment Time	8	7	5	20
Promotion					
17.	Advertising	4	1	15	20
18.	Personnel selling	5	15	–	20
19.	Direct Approaches	6	12	2	20
20.	Distributorship	1	1	18	20

Source : Computed from Primary Data.

packaging. With regard to quality, nearly 15 enterprises are concentrated. Only one enterprise is concentrated on quoted price and nearly 18 industry more concentrates in credit period and payment time. Another 15 to 18 enterprises are having less concentration in advertising and distributorship, 12-15 enterprises concentrates in personal selling and direct approaches.

Gathering marketing information through market survey by being around people, observing them and asking them question by way of direct contacts may be possible for a small enterprises having limited market network and customers constellation. But on the face of this age of market economy followed by globalisation. Even a very small organisation has the need for much more and better information about market, gradually expending and dispersing all over. Hence adequate information an competitors, resellers and other interacting forces influencing customers on the global range has become imperative in analyzing marketing opportunities through a streamlined market survey.

Table 15.2 : Market Survey in Micro-Enterprises

	No. of Micro-Enterprises		
Stages of Market Survey	*Adopted*	*Not Adopted*	*Total*
I. Collection of Primary Information	15	5	20
(a) Non-reactive method	3	–	–
(b) Reactive method	2	–	–
II. Compilation and Analysis	4	16	20
(a) Government Publication	–	–	–
(b) Books and Periodicals	2	–	–
(c) Published Data	2	–	–
(d) International Information	–	–	–
III. Applicability	2	18	20

Source : Computed from Primary Data.

Table 15.2 shows the market survey stages followed by the sample interprises. It is clear that 15 micro enterprises adopted the collection of primary information mostly under reactive method. Most of enterprises (16 enterprises) did not adopted the compilation and 18 enterprises are not adopted analysis techniques with regard to market survey applicability stages.

Problems and Suggestions

The micro-enterprises have been, facing stiff competition from the modern manufacturing sector. Their survival has been asked only because:

- There was partial stage support.
- Some products could be marketed in places where the large scale sector products did not make an entry owing to wide geographic spread of village habitat.
- Production led approach being adopted as against a market oriented approach resulting in stock pile up, obsolete products, precarious liquidity position and reduced profitability.
- Lack of information on market research findings, especially with regard to consumer preference and absorption capacity of markets.
- Difficulty in access to information regarding legislation regulation pertaining to marketing of products (especially exports).
- Poor presentation and packaging of products and limited opportunities for advertising largely due to low margin and Shoestring budgets for product promotion.
- Lack of knowledge and information on suitable/appropriate marketing channels and distribution networks.
- Lack of access to international markets, especially with regard to consumer preference in overseas markets, overseas regulations and competition within and outside India.
- Competitions from the unorganized private sectors (with low labour costs), the organized private sector (sophisticated, capital intensive production facilities with mass production systems backed by strong research and advertising push) and the regulated Government sector (surviving on liberal production subsidies, heavy seasonal discounts on sales and baurecratic/political support).
- Unlike some western countries, in the absence of a charity market, a consumer in urban market does not attach any additional value to a product being marketed by rural people agency and to that extent, such product is denied the advantage of a USP (Unique Selling Proposition), UNICEF and CRY have been successful in using this USP in India for their greeting cards.

The suggested marketing strategies for he Micro Enterprises are :

1. *Local Rural Markets:* These could play a vital role in ensuring cash flow, stability, growth and influencing selection on non-farm options, with growing rural markets. It is increasingly being felt that their capacity to absorb products and service is increasing. They offer a potential for direct marketing by producers where poor have better access and control in comparison to urban market which are highly competitive and full with various dynamics. It is also desireable to explore local markets as this interaction is locally manageable and easily undestood.
2. *Market Intelligence and Forecasting Models :* Market intelligence reports on various types of products are needed so that the producers benefit in procurement, storage and marketing of such products. Future trading on commodity sector is on the anvil. Designing forecasting model and forecasting product demand/supply including price shall be of great advantage for the poor producers.
3. *Quality Control and Grading:* Quality control and grading plays an important role in sale of rural products, determination of price and consistent demand from the producers and sellers. The strategic intervention in this direction could be :

   (a) Making the primary producers/primary NTFP collectors conscious of quality and preservation/ maintenance techniques.

   (b) Facilitating producers in procuring quality certificate for their products (Agmark, ISO 9000 and 9001) .

   (c) Advocating introduction of grading/quality control at the procurement/collection. For example, in Orissa, the primary collector in Kendu leaf receive price irrespective of the grade of the leaf even though Bolangir Kendu leaf have high export demand for the countries like Bangladesh and Pakistan.

4. *Distribution*: The aim of distribution should be to reduce trade channels as far as possible. The strategic intervention in this regard could be:

   (a) Forward intervention to encourage producer groups to go forward into starting its own distribution system from mere manufacturing. This will help them gain independence with regard to retail outlet.
   (b) Promote creation of local trade agent who can replace exploitative trade agents.
   (c) Encourage women SHGs of the locality to start retailing/vending in the local markets to gain more access on the market.
   (d) Encourage producers to sell handicraft items/ consumer products through wholesale distributors and retailers. The current perception of skimming pricing (the initial prices set are at a higher level and the objective is to skim the market) strategy should be changed.

5. *Design Development:* Design development should be undertaken with following objectives:

   (a) Diversifyng the design of products,
   (b) Enhancing the range of products,
   (c) Substitution of products (using local raw materials),
   (d) Enhancing utility and user friendliness of products,
   (e) Value addition of the products,
   (f) Retaining the ethnicity and traditional art through workshops and revitalization of dying craft.

6. *Marketing Infrastructure* : Provision of marketing infrastructure in these areas can to a long way enable the rural poor to market their goods and at the same time building an identity for the products. The self help groups

can be encouraged to try out these markets on their own so that they slowly develop marketing skills. They can also start marketing the goods of the individual swarozgaris. Infrastructure fund can be used for strengthening marketing infrastructure and related activities.

7. *Intermediary Agency :* The agencies can be in the governmental, non-governmental or private sector. The one aspect that requires to be taken care of is the stability of the agency, their own competence and their capacity to ensure that the rural producer is getting his/her fair share of the profits. Care should be taken to see that only such agencies that do not exploit the rural producers and are well established in the business are chosen for the purpose. These agencies may market the products in the national or international markets.
8. *District Supply and Marketing Society (DSMS) :* DSMS must provide non-credit inputs like procurement or purchase or raw materials and also for sale of products. Wherever these societies are doing good work, they can be suitably strengthened. Similarly, in states such as Orissa and Kerala, there are marketing societies at the State Level. Their capacity and product range can be suitably expanded. However, it is not always necessary to establish new agencies. It should be a felt need after a careful analysis of the existing infrastructure, and should be part of a strategy for promotion of marketing support, to products under SGSY.
9. *Khadi and Village Industries Commission:* The marketing outlets of KVIC/KVIB must be made use of for SGSY products. Keeping the market trends in view, linkages with apex bodies can ensure a stable market at remunerative prices. Such a link will also result in quality improvements while reducing the market risks to the poor. Bankers can also encourage such initiatives through liberal craft delivery since the loan recovery can be built into the project. There are several NGOs and other independent institutions to

market products from rural artisans/craftsmen across the country. DRDA should strive to develop linkages with such groups.

10. *Publicity :* A few states have organized to give publicity to the range of products produced by the Swarozgaris. Such efforts need further encouragement so that SGSY Swarozgaris will be integrated with the regional markets. Organizing of exhibitions periodically in urban centres will offer better price to the goods of SGSY Swarozgaris. This also serves as forum for sensitizing the SGSY group to the demand patterns of the market. The state level organizations and boards have to develop as organic link with the SGSY groups and provide virtual linkage. Thus, professional bodies can guide the poor in supply of designs and in quality assurance. Attempts to establish state emporia for display and sales of SGSY products in major urban cities need to be given a fair trial.
11. *Quality of Products:* The quality products is vital not only for the development of entrepreneurs but also for the nation as a whole. It must be the responsibility of the marketing agencies to ensure that the Swarozgaris are trained in quality control. The goods marketed must conform to a minimum and uniform quality so as to be able to develop a brand image. This is important even in respect of the goods that are produced for local consumption. The DRDAs must organize periodic meets of the Swarozgaris and ensure that they are given the necessary guidance in quality control. There must also be a system of the experts in the relevant fields visiting the work places and guiding the swarozgaris.

Conclusion

Sustainability of micro-enterprises as a group activity under SGSY is a difficult proposition unless supported by a committed and capable NGO. It is easier to form SHGs and sustain them as savings and credit groups, but quite a challenge to develop at as a comprehensive support institution for micro-enterprise development. In Dindigul District self help promoting institution implementing SGSY have not acquired the required capability to promote micro-enterprise development is a significant way. But the SHGs they promoted have facilitated capital formation and credit flow. Though many NGOs came forward to form SHGs and assist them to establish linkages with banks, several of them shy away in so far as promotion of micro-enterprises is concerned. Hence, marketing support for SGSY product needs to be strengthened in the districts through bazaars retailing, wholesale, export from the SHG's products through their chains of shops.

REFERENCES

Books

KG. Karmaker, *Rural Credit and Self Help Groups*, Sterling Publishers, New Delhi, 1999.

N. Litha and B.S. Nagarajan, *Self Help Groups in Rural Development,* Dominant Publisher and Distributors, New Delhi, 2002.

M. Soundarapandian, *Rural Industrialization Problems and Prospects*, Anmol Publication, New Delhi, 2003.

M.Soundarapandian, *Rural Entrepreneurship*, Kanishka Publication, New Delhi, 2000.

Asghari Mohiuddin *et al.,* "Access to credit for women beneficiaries under SGSY", NIRD Research Highlights - 2001-02, National Institute of Rural Development, Hydrabad, 2 003, pp. 161-172.

Journals

K.K. Tripathy, "Self-Help Groups—A Catalyst of Rural Development", *Kurukshetra*, Vol. No. XI, June 2004, pp. 40-43.

P. Purushotham, "Marketing support to SHGs", *Kurukshetra*, Vol. XI, No.2, February 2004, pp. 23-28.

K. Surekha Rao and G. Padmaja, Self Help Groups in Tripathi, A.P., *Social Welfare*, Vol. 45, No. 1, April 1998, pp. 25-29.

Punitha, M. Women Entrepreneurs : Their problems and constraints, *The Indian Journal of Labour Economics* 1999, Vol. 42, No. 4, pp. 707-709.

Manimekalai, N. and Rajeswari, G. Empowerment of women through self-help groups", *Margin* Vol. 32, No. 4, July-September 2000, pp. 74-87.

Datta, K. Samar and M. Raman, "Can Heterogeneity and social cohesion co-exist in SHGs : an evidence from group landing in Andhra Pradesh to India". *Indian Journal of Agricultural Economics* Vol. 156; No.3, July-September 2001, pp. 387-400.

M. Punitha. "Women Entrepreneurs, their. problems and constraints", *The Indian Journal of Labour Economics,* 1999; Vol. 42, No. 4, pp. 707-709.

Manimekalai, N. and Rajeswari, G. "Empowerment of women through self-help groups", *Margin* Vol. 32, No.4, July-September 2000, pp. 74-87.

Archana Sinha, "Micro Finance for Women ·Empowerment—a perspective" *Kurukshetra*, Vol. 52, No.6, April 2004, pp. 31-38.

16

SUSTAINABILITY OF SSI DURING ECONOMIC REFORMS ERA

M. Soundarapandian and *A. Chandraprabha*

Introduction

The current year has witnessed the dramatic collapse of many a South East Asian economy. Indonesia, Malaysia, Thailand, South Korea were once considered shining examples of economics on the upswing, which wrought miracles. Japan, a major creditor nation, which could have bailed out its Asian neighbours, has itself become part of the regions current problems. The Japanese economy has become so enervated that it cannot act as a catalyst to boost the regions sagging fortune. The crisis has manifested in the forms of sharp depreciation of the local currencies, inability to service sovereign debts, mounting bad debts of banks, bankruptcies of corporates, steep decline in stock market indices, etc.

That the Indian economy remained almost unaffected by the South East Asian crisis is a tribute to its strong fundamentals. India opted for a restrained and prudential liberalization path in the Eighties and took to economic liberalization vigorously in the Nineties. Reforms in the banking sector, while ringing in wide ranging de-regulation, introduced prudential asset classification and income recognition norms. Disclosure norms were made stringent. These measures considerably strengthened the banking system and at the same time isolated the weak banks, the problems relating to which could be tackled separately.

The capital market was also liberalized which enabled corporates to tap public money to meet their requirements without many hassles. The Securities and Exchange Board of India, the regulatory authority for capital markets, was set up and made autonomous. Foreign Institutional Investors were allowed to invest in Indian companies within an overall ceiling.

The small scale industrial sector which plays a pivotal role in the Indian economy in terms of employment and growth has recorded a high rate of growth since Independence in spite of stiff competition from the large sector and not so encouraging support from the Government. This is evidenced by the number of registered units which went up from 16,000 in 1950 to 36,000 units in 1961 and to 33.7 lakh units in 2000-01. During the last decade alone, the small-scale sector has progressed from the production of simple consumer goods to the manufacture of many sophisticated and precision products like electronics control systems, micro-wave components, electro-medical equipment, T.V. sets, etc.

The Government has been following a policy of reservation of items for exclusive development in the small-scale sector. At the time of the 1972 census of small-scale Industrial Units, there were 177 items in the reserved list. By 1989, the reserved list excluded 837 items for exclusive production in the small-scale sector. These units produce over 7,500 commodities.

Economic Reforms of SSI

The following are the economic reforms aimed at economic development:

Industry and Service

(a) The Information Technology Bill to create the legal framework for facilitating electronics commerce in the country was introduced in the winter session of the parliament.

(b) Tax provisions for housing have been liberalized and the banks have been permitted to land up to 3 per cent of incremental deposists for housing.

Infrastructure

(a) New telecom policy allows multiple fixed service operators and opens domestic long distance services to private operators. It also allows existing licenceholders of basic and cellular services to migrate to revenue sharing arrangements.

(b) TRAI reconstituted through and ordinance, which envisages clear distinction between the recommendatory and regulatory functions of the authority. It will be mandatory for the Government to seek TRAI's advice on policy and licensing issues. A separate Appellate Tribunal will hear appeals against the decisions of TRAI and the Government licensor.

(c) Restructuring of airport through long-term leasing route.

(d) Indian Railway Catering and Tourism Corporation (IRTC) has been incorporated as a Government company with the objective of upgrading and managing rail catering and hospitality.

External Sector

(a) Foreign Exchange Management Act (FEMA), 1999 has been enacted. The new Act replaces the old FERA. Its provisions are in conformity with a liberalized market in foreign exchange.

(b) Removal of Quantitative Restrictions has been made. Import of 894 items has been made licence free and another 414 items permitted to be imported against SIL in EXIM policy change announced on 31 March, 1999. The Government has further decided to phase out all QRS maintained on BOP grounds by April 1, 2001.

(c) A new chapter in EXIM policy to boost export of services has been incorporated.

(d) Free Trade Zones (FTZ) to replace Export Processing Zones and these are to be treated to be outside the country's customs territory.

(e) Except for negative list, sectoral limit, and a few explicitly defined constraints, all other FDIs will now be under the RBI's automatic system. NRIs/OCBs have been permitted to invest under the automatic route in all terms, baring a few.

(f) Indian companies are free to access the ADR/GDR markets through an automatic route subject to the specified norms and post-issue reporting requirements. Such issues would, however, needs to conform to the existing FDI policy.

Financial Sector

(a) The Insurance Regulatory and Development Act (IRDA) has been passed by the Parliament in December 1999, which seeks to promote private sector participation in the insurance sector and permits foreign equity stake in domestic private insurance companies up to a maximum of 26 per cent of the total paid up capital.

(b) Ordinance has been issued to amend the recovery of dues to the Banks and Financial Institutions Act, 1993. The amendment strengthens provision for recovery of dues owed to the Banks and the Financial Institutions.

(c) Banks allowed operation different Prime Lending Rates (PLRs) for different maturities.

Capital Market

(a) The Securities Laws (Amendment) Bill, 1999 has been passed by the Parliament incorporating derivatives and units of Collative Investment Schemes (CIS) in the definition of securities in the Securities Contract Regulation Act, 1956. CIS regulation have been notified in October 1999.

(b) Announcement of calendar for issue of Treasury Bills and re-introduction of 182 days Treasury Bills has been made.

(c) Rolling settlement has been introduced in 10 select scrips.

(d) Companies have been given free to determine par value of shares issued by them.

Taxation

(a) Major Reforms of Central Excise have been made. Eleven major *ad-valorem* rates of excise duty have been reduced to three, namely a central rate of 6 per cent, a merit rate of 8 per cent and a demerit rate of 24 per cent.
(b) The cap on MODVAT credit of 95 per cent of the admissible amount has been lifted and restored to 100 per cent.
(c) Reduction in protective custom tariff from 45 per cent to 40 per cent.

Sustainability of SSI

A significant features of the Indian economy since Independence is the rapid growth of the small industry sector. In the Industrial Policy Resolution of 1948 and 1956, the small sector was given special role for creating additional employment with low capital investment. A new thrust was given in favour of small units by the Industrial Policy Statement of 1977.* In 1950, the government grouped Small Scale Industrial Undertakings into two categories those not using power but employing less than 100 persons. All small-scale enterprises, however, had capital investment of less than Rs.5 lakhs. None of these criteria taken singly would be a determining rest as they undergo changes over a period of time. The third criterion, namely the character of organization and management, also cannot be considered a sound basis of classification. Apparently, the standing feature of small-scale enterprises seems to be the personal character of its organization and management in contrast with the predominantly impersonal organization and management is predominantly proprietary with individual ownership or partnership. But the ownership and management may also be identical in some of the large scale industries. The criterion, therefore, becomes vegue an inappropriate.

In 1966, the small-scale enterprises were defined as

* Yogendra Nath Mann "Second Generation Economic Reforms" edited by K. Sham Bhat "Indian Economy under Globalization Process" serials publication, New Delhi, pp. 437-442.

undertakings, with fixed capital investment of less than Rs.7.5 lakhs and ancillaries with fixed capital investment of Rs.10 lakhs. Investment will imply investment in fixed assets in plant and machinery, whether held in ownership term or by lease or by hire purchase. In 1975 this limit was revised to Rs. 10 lakhs for small-scale enterprises and Rs. 20 lakhs in case of ancillaries. Subsequently, under the Industrial Policy Statement of 1980 this limit was further raised to Rs. 20 lakhs in case of small-scale units and Rs. 25 lakhs in case of ancillary units. Simultaneously, in the case of tiny units, the limit of investment has been raised from Rs.1 lakh to Rs.2 lakhs. In March 1985, the Government has again revised the investment limit of small-scale to Rs.35 lakhs and for ancillary units to Rs.45 lakhs.

As per the industrial policy statement of May 1990 the investment ceiling in plant and machinery for small-scale industries (fixed in 1985) has been raised from Rs.35 lakhs to Rs.60 lakhs and correspondingly for ancillary units from Rs.45 lakhs to Rs.75 lakhs. Investment ceiling with respect to tiny units has been increased from Rs.2 lakhs to Rs.5 lakhs. According to the noted definition an ancillary unit is one which sells not less than 50 per cent of its manufactures to one or more industrial units.

During 1997 on the recommendation of Abid Hussain Committee. The Government has raised the investment limit on plant and machinery for small units and ancillaries from Rs.60.175 lakhs to Rs.3 crores and that for tiny units from Rs.5 lakhs to Rs.25 lakhs.

The Government in 2000 has reduced the investment limit on plant and machinery from Rs.3 crores to Rs.1 crore, but the limit for investment in tiny units has been retained as Rs.25 lakhs.

Policy Interventions

The Government of India has always been highly supportive of the SSI sector. A large number of fiscal and other types of concessions/ incentives have been made available to the sector support policies have undergone several changes in tune with the scenario changes:

(i) Revision in ceiling on investment in plant and machinery in SSI units from Rs.3.5 million to Rs.6.5 million in 1991 and very recently to Rs.30 million in 1997. This is primarily intended to facilitate acquisition of better technologies on concessional terms.

(ii) Large industries permitted to invest inequity of SSI up to 24 per cent.

(iii) Indirect taxes and import levies have been significantly moderated from import duties of over 100 per cent, duties now are as low as 5 per cent on some inputs.

(iv) Capital goods/technology imports have been made easier.

(v) Indian investment in joint ventures abroad has been liberalized.

(vi) SSI sector is exempt from payment of excise duty up to a turnover of Rs.10 million.

The provision variable to large scale industries to take up equality upto 24 per cent in SSI units is a major step which leads to a win-win situation for both. The large industries can participate in the equity of ancillaries and intermediate product manufacturers and bring to beer their technological and managerial capabilities on the running of such SSI units.

Problems of SSI

The problems generally faced by small-scale entrepreneurs in India in the process of setting up running units can be summed up under the following heads :

Finance

Large units raise their finances by public subscription, equity and other securities, which are available for major, capital investment without repayment liability and interest over and above this, an investment allowance, development rebate, depreciation etc., are also permitted which help in developing a sound financial base by generating such reserves.

Land Building

Non-availability of land buildings particularly in industrial estates is another problem faced by the small-scale entrepreneurship even where these are available the premises are often unsatisfactory and longes fed.

Labour

The problems faced by the small-scale units which required to labour can be summed up to under the following heads:

(a) Non-availability of skilled workers,
(b) Weigh high real cost of labour, and
(c) Insecurity arising out of an estrus of labour to large firms.

Marketing

While the number of small-scale units and their production has gone up it has come to light that marketing of the products is creating a number of problems, which can broadly be classified under the following heads:

(a) Inability to compete with large-scale sector manufacturing similar products for a wider consumer market;
(b) Lack of guidance to reach conventional market channel and establish new channels for marketing;
(c) Inability of small-scale units to reach export market on their own due to lack of adequate resources.

Raw Materials

There is lack of adequate quantities and of good quality raw materials. The most important raw material for engineering industry in the small-scale sector are iron and steel and non-ferrous metals, with regard to supply of iron and steel our inquiries reveal that the system of distribution is defective and the quantities earmarked are also inadequate for this sector units have hardly been given 10 tonnes of iron or steel in a period of six months with this meagre to negative these units can't function economically.

Moderation Quality Control and Standardization

Small-Scale Industrial Units in the engineering industry face problems of modernization, quality control and standardization the engineering industry has immense potential to produce a variety of goods but is unable to keep abreast with product design, technological changes because of the above mentioned handicap.

Power Failures

Frequent power failures and cuts are also affecting the efficiency of small-scale units. Irregular power supply prevents full utilization of capacity and raises cost of production. Their individual power consumption is small. Therefore, captive generation of power by installing their own generation sets in uneconomical.

Strategies

The Government may adopt the following suitable strategies for the development of SSI under the liberalization era.

- The credit limit under the credit guarantee fund scheme may be enhanced to Rs.50 lakhs per borrower of SSI units;
- The excise exemption limit can be increased to Rs. 2 crores for small-scale entrepreneurs;
- The scheme of Technology Upgradation and Management Programme (UPTECH) is to be implemented to the viable small-scale industrial units. The awareness may be created about the credit linked capital subsidy scheme of the Government of India;
- The Government should create Industrial and Science Parks with best possible infrastructure.

Strengths

- As SSIs operate in the narrow band product range, the thrust on customer satisfaction is paramount by way of quality service, and price;
- Being individual or small group driven, bereft of rigid hierarchical layers, these units are more flexible;
- With their ears to the ground, these units can anticipate, and readily respond to change;
- Compared with medium or large scale units, operating costs by way of overheads, materials, administrative expenses, etc., are lower.

Weaknesses

- Promoters alone take decisions. The possibility of taking an erroneous decision is high;
- Individuals take pence over a cohesive system;
- Reasonable progress satiates the promoter's desires;
- Being individual or family owned, genealogical succession is a fact. This may lead to incompetence, sickness, or demise of units;
- Attracting and retaining outside talent is a problem.

The Future in the year 2010

The participants forecast the emergence of the following trends:

- Strengths remain the same as in 1996, if not enhanced;
- Automation levels will increase considerably travel and communication will be easier, and aid access to national and international technologies;
- A customer driven economy will compel medium and large industries to offload more manufacturing and services to the SSIs;
- Create emphasis will be placed on product quality and perfection. Total Quality Management (TQM) will play

an important role, and SSIs should be vigilant about this aspect. Otherwise, existence will not be possible.

- There will be an exponential growth in productivity by the application of advanced technology. This would reduce costs and increase the margin of returns;
- Virtual abolition of trade barriers, easy availability of technology and equipment, thinning of boundaries between small and medium industries;
- Simplified/suitable labour laws to respond to emerging flexible, efficient, and committed labour force that is capable of maintaining the rapid pace of production with quality;
- Rapid technological obsolescence, faster than in the 1980s and 1990s exerting tremendous pressure on SSIs to upgrade/modify product/ process technologies;
- Government grants, subsidies, reservations, etc., will disappear and the small industry sector will react to market forces more sharply;
- Inventories will come down to optimum levels concepts like just in time will emerge as the cost of storage will become less affordable.

The trends in infrastructure have been perceived as follows:

- More industrial/technology parks with all facilities for small industry will be established;
- Availability of uninterrupted, quality power;
- Simplified, market-driven export incentives;
- Simplified, speedy clearances to setup new industries.

The number of SSI units will group:

- Cooperative arrangements between units manufacturing the same product to consolidate negotiating power in the market.

The likely scenarios that will emerge in the future may be as follows:

- To be able to alter large production capacities which will interest domestic and international buyers, conglomerates of like minded and like product units will operate without technically enlarging their six;
- Quality will be the uppermost concern in a customer-oriented economy ISO 9000, other reigning and emerging standards will become important. This will increase the competitiveness and the confidence of small units;
- All operations will be electronic and sophisticated technology would be adopted at all levels;
- Small units will be environment friendly. Necessary certifications will be obtained;
- As transportation costs will increase units will be located or will shift closer to markets;
- Small of today will become closer to medium of tomorrow, and together they would be in position to play a major role in the country's economy.

The tiny sector programme content has been envisioned as follows:

- Greater weightage to tiny units in the programme content of the SSI sector;
- Marketing applicable integrate term loan upto Rs.0.5 million on preferential terms compared with bigger SSIs;
- Easily implementable excise concessions and graded interest rate structure for loans on hire purchase of machinery;
- Simplification of rules and procedures, and removal of all clearance hurdler;
- Preferential treatment in allotment of infrastructural facilities and development of mini industrial estates for tiny sector units;
- Concentrated efforts of industrial extension services towards technology upgadation and modernization;

Table 16.1 : Performance of SSI Sector

Year	Number of units (Nos. in lakhs)			Fixed Investment	(Rs. in crores)		Employment (No. in lakh)	Export Rs. crores
	Regd	In Regd	Total		Production (at current prices)	(at constrant price of 1993-94)		
1990-91	NA	NA	19.4	NA	155340	NA	125	9763
1991-92	NA	NA	20.82	NA	178699	NA	130	13883
1992-93	16.48	5.98	22.46	NA	209300	225626	134.06	17784
1993-94	17.76	6.12	23.88.	NA	241648	241648	13938	25307
			(6.14)			(7.10)	(3.97)	
1994-95	19.44	627	25.71	NA	298886	266054	146.56	29068
			(7.66)		(23.69)	(10.10)	(5.15)	
1995-96	20.18	6.40	26.58	NA	362656	296385	152.61	36470
	(3.81)	(2.07)	(3.88)		(21.34)	(11.40)	(4.13)	(25.46)
1996-97	21.82	6.50	2.03	64839	411858	329935	160.00	39248
	(6.69)	(1.56)	(5.46)		(13.57)	(11.32)	(4.84)	(7.61)
1997-98	22.82	6.62	29.44	70891	462641	357749	167.20	44442
	(5.59)	(1.85)	(5.03)		(12.33)	(8.43)	(4.5)	(13.23)
1998-99	24.06	6.74	30.80	77531	520650	385296	171.58	48979
	(5.43)	(1.81)	(4.62)		(12.54)	(7.7)	(2.62)	
1999-00	25.26	6.86	32.12	82411	572887	416736	178.5	54200
	(4.99)	(1.78)	(4.29)		(10.03)	(8.16)		
2000-01	26.14	6.98	33.12	89275	639024	478456	192.23	NA
	(5.78)	(1.750)	(4.92)		(11.54)	(6.08)	(3.9)	
2001-02	27.31	7.11	34.42	NA	670316	451033	192.23	NA
			(4.92)		(8.03)	(8.23)	(3.5)	
2002-03	28.49	7.23	35.72	NA	742021	514292	199.65	NA
			(3.8)		(7.5)	(7.5)	(3.9)	

Source: Annual Reports, Ministry of SSI, New Delhi, 2002-03.

- Closer monitoring of SSI units to minimize sickness and mortality.

Conclusion

Globalization and change in technology are emerging as the major forces that will modify and mould the environment for small business and entrepreneurship over the next decade which will witness the emergence of more competitive and efficient SSI sector. The role of the Government will be that of a facilitators for providing support to SSI in the era of Globalization. The Government would increasingly focus its energies towards providing human resources development support along with establishing quality certification facilities and dissemination of information.

REFERENCES

Reports

1. Annual Reports, Ministry of Small-Scale Industries, Government of India, 2002-03.
2. P.B. Nimbalkar, "Small and Medium Enterprises Emerging Issues", Laghu Udyog (Small-Scale Industries), Ministry of Small-Scale Industries, Government of India, Nirman Bhavan, New Delhi 2002, pp.10-18.
3. Inter University Seminar on Small-Scale Industries, March 18, 2004.
4. Government of India, Ministry of Small-Scale Industries, Annual Report, 2001-02.

Books

1. Ruddar Datt, K.P.M Sundharam, "Indian Economy", S. Chand & Company Ltd., New Delhi, 2004, pp. 234-262.
2. Vijay Joshi, IMD Little "India's Economic Reforms 1991-01", Oxford University Press, Delhi, 1997, pp. 63-107, 171-211.
3. R.P. Singh, "New Economic Policy and Perspective to Rural Development", National Institute of Rural Development, Rajendranagar, Hyderabad, 1995, pp. 53-60.
4. R.C. Choudhury *et al.*, "Micro-Credit for Micro-Enterprises", National Institute of Rural Development, Rajendranagar, Hyderabad, 2001, pp. 89-99.

Journals

1. M.H. Bala Subramanya, "*Small Industry and Globalisation—Implications,* Performance and Prospects", *Economic and Political Weekly*, May 1, 2004, Vol XXXIX, No.18, pp.1826-33.
2. R.N. Sengupta, "Preparing SMEs in India to Meet New Challenges, *SBI Monthly Review*, November, 1998, pp. 578-580.
3. Nanjundappa, "Small-Scale Industry in the Globalisation Era", *Southern Economist*, Vol.42, No.9, July 15, 2002, pp. 5-8.

17

A NEW APPROACH TO HUMAN RESOURCE DEVELOPMENT AND AGRICULTURAL PRODUCTIVITY—A CASE STUDY

S.N. Sukumar

Introduction

Higher income and more output, by and large, characterize economic development. It is referred to higher national income and national output. This condition indicates increased purchasing power of the people and productive potential of the economy. This can be achieved by increasing the level of input. The increased input may be in terms of physical capital or human capital. Increased physical capital may be realized in terms of capital or exploitation of natural resources. However output can be increased without increasing the financial commitment or capital requirement. By improving the quality of labour force through proper training programme we can improve their performance and production capacity of the economy

Indigeneous Development

Emerging Development within the system can be sustained forever. Development through external factors will not long last for ever. Development through external factors either lead to foreign dependence or damaging the environment. The so-called modern technology and industrial development make people struggle to get fresh air and food. The strategy of mass production drives out the mass from the development orbit. The people should accomplish the new development.

New Approach to New Development

Economic development should be preceded by people's development. Over all development or the total development should be preceded by rural development. The development path should not march towards villages from city rather it should march towards city from villages, that is through decentralization. Centralization leaves the people around the development. Development through rural development can be sustained for ever as it is by the people and for the people.

Rural Human Capital Formation

The significance of human capital is implied in the concept of organization as one of the most important factors of production. Human capital refers to the organizing ability and innovative ability of the people. Qualitatively elevated labour force is the form of human capital. Today, the concept of human capital may be classified into Rural Human Capital and Urban Human Capital. It should be further classified into habitual and educational. Educational means must be further classified into formal and informal education. Habitual approach is much more related to psychological aspects of mankind.

Educational approach improves the existing knowledge of the people, which in turn improve the efficiency of the people. The knowledge of the society is being improved through schools, colleges and higher education. The education is imparted through formal and informal means. In formal education is being imparted through adult education in general and Agricultural Extension education and training in particular to agricultural sector. Psychological oriented informal education much more effective than the formal education. Oxford University, in its survey, revealed that the success of the individual was determined by the knowledge gained out side the schools and colleges. The knowledge equipped in the schools and colleges helped only 15 per cent in their success.

The knowledge developed outside the schools and colleges are referred to leadership quality, personality development, motivation, self-confidence, creativity, positive attitude, self-esteem, etc. These are the qualities not developed in the schools and colleges. These subjects are not included in the regular curriculum of the schools and college studies. But these qualities have more impact on individual success rather than the usual schools and college education. It is worth to mention here that Galbraith, the popular economist, who stated in his book mass poverty that dishonesty, easy going attitude and lack of trustworthy are responsible for India's backwardness. These are merely mind-related or psychological concepts not economic concept.

At present, therefore, what we need to increase our income and output is not more capital or investment and technology but change in our attitude, approach and outlook. We need changes in our intra and inter-relationship with our co-livers. What we have to understand or learn is to practice what we know. We know many things but we never do it. We have more words and not deeds. Let us motivate people to do what they know.

Habit of saving among the poorest of the poor must be encouraged. Each one in the rural economy must be known to banking formalities and procedure. At least 5 per cent of their income must be saved and there should be bank account for all the people. Let them deposit their surplus, if any, in the bank.

They should avoid taking loan for unproductive consumption. They should spend out of their income not borrowings. They should have faith in themselves and in their work. They should work and earn. Attitude of expecting their bread from others or from the government should be avoided. Significance of self-help concept must be taught among the people. Through this concept of self-help, indirectly we help others. Helping others is the purpose of life. This is achieved through self-help.

The terms of trade between rural and urban must be favourable to rural people so that rural urban migration may be avoided. Life in the village should make them to stay in the village itself. The opportunities and the facilities should go to villages instead of rural people going to towns and cities. One has to be Indian in spirit and values.

Let them do away with the fatalistic attitude. Make them optimists. Let us evolve a new plan to make village people to live with vigor and vitality, strength and stamina, courage and conviction, patience and perseverance and creativity and productivity. Let them not loose their identity in the urban ocean. For example, let us see the impact of agricultural extension education and training on farmers productivity. In the Sozhamore Panchayat, Vellore District. The case study explains that the farmers should be educated about cultivation regarding importance of cropping pattern, cropping intensity and new farming technology. The study reveals that the agricultural productivity is more among the farmers who got the information and guidance from agricultural officers.

Case Study

Agriculture in many underdeveloped countries has the potential capacity to produce adequate food for the steadily growing population and also to improve income and welfare of the poor people significantly. Agriculture is the first science, the model science and the mother of all sciences. The success or failure of agriculture determines the success or failure of sciences as a whole. Infact, agriculture feeds the mankind and supplies raw materials for agro-based industries, which of course facilitates the process of development in the low-income countries. It is the matter of survival for the mankind and the animals. Schultz emphasized that a well-developed agricultural economy is an essential element for balanced growth.

Therefore, agricultural development in poor countries requires shifts from less productive to more productive techniques per worker, sale of produce in the organized markets, creation of new skills, and accumulation of new knowledge. The accumulated knowledge constitutes the true capital of a country, which in turn gives opportunities for invention, innovation, and transformation of agrarian economy into industrial economy. The benefits of such knowledge can be made available to society through Agricultural

Extension Education and Training Programme. The transmission of accumulated knowledge enriches the human resources for better performance in agriculture. Hence, the researcher made an attempt to study the impact of Extension Education and Training on the performance of farmers in terms of farm input-use-efficiency, productivity and profit rate among the contact and non-contact farmers in Tamil Nadu with special reference to Sozhamur Panchayat, Vellore district. Farmers cultivating under the supervision of Agricultural Extension Officer are known as contact farmers and farmers not under the guidance of Agricultural extension officer are the non-contact farmers.

Objective and Hypothesis

The study has been made with the broad objectives to find out, farm-input-use-efficiency, productivity and profit rate for contact and non-contact farmers.

To fulfil the objectives mentioned above the following hypotheses have been formulated and tested.

The contact farmers use inputs more efficiently than the non-contact farmers.

The yield rate and the rate of profit are greater among the contact farmers than the non-contact farmers.

Analytical Framework

Productivity is generally defined as the output per unit of input. It is the ratio between output and input. The earliest approach to the measurement of productivity was based on the reciprocal of capital output ratio. In 1960s, Solow and Kendrick estimated total factor productivity growth based on the Cobb-Douglas production function. More recently May and Denny (1979) and Chen and Tang (1990) and host of others utilised the translog production function to estimate the Divisia index total factor productivity growth. Similarly, World Bank used the Gollop and Jorgensen model (1986) to measure TFP in selected developing countries. There are several approaches to study the impact of extension education and training on farm production. Feder, Gerson and Roger H. Slade (1987) used cost-benefit analysis to evaluate the impact of training and visit system on farm return. Huffman, Wallace (1972) compared the expenditure on research and extension education on crop productivity to examine the effectiveness of extension education. Benore and Harrison (1977) compared the yield rate between the areas with and without extension services. Kalirajan, K.P. and Shanol, R.T. (1984) made an attempt to compare the farmers with the exposure of extension education and training with farmers without the exposure of extension education. In this work the authors used multiple linear regression model and capital-output ratio to estimate input-use-efficiency in agricultural sector between the contact and non-contact farmers. In addition to that, Joan Robinson growth model also was used to examine the input-use-efficiency in terms of profit rate among the contact and non-contact farmers.

Methodology

In order to estimate the level of productivity among the contact and non-contact farmers personal interviews, informal discussion, participant and non-participant observation have been adopted. The informations were collected from seven villages of Sozhamur Panchayat. As per the 1991 census there were 750 farmers in the study area, of which 70 farmers are contact farmers and the rest 680 farmers are non-contact farmers.

In the case of comparative analysis, the methodology is very important. By and large, the number of impact studies conducted with the comparisons between beneficiaries and non-beneficiaries. However, farmers are different in several aspects such as socio-economic conditions, factor endowments of the region and other infrastructural facilities. It is, therefore, necessary to classify the farmers with similar circumstances. The pilot survey indicated that all the seventy contact farmers cultivate more than one acre, irrigate crops by pump set and cultivate high yielding variety of paddy, viz. ADT36.

Therefore, the researcher selected 115 non-contact farmers out of 680 non-contact farmers who comply with the above mentioned conditions. Thus the sample participants are the different category of farmers, viz. contact and non-contact farmers but they are cultivating in the similar environment.

Techniques of Analysis

To determine the efficiency of factors of production in farming, multiple regression analysis, reciprocal of capital output ratio and the rate of profit (Joan Robinson's growth model) were used.

The regression model used was of the form

$$Y= \beta_0+\beta_1X_1+\beta_2X_2+\beta_3X_3+\beta_4X_4+\beta_5X_5+\beta_6\times 6+U$$

where

Y	=	Yield (in kg) per acre
X_1	=	Send cost (in kg) per acre
X_2	=	Size of holding in acres
X_3	=	Labour cost in mandays per acre
X_4	=	Fertilizer cost per acre in rupees
X_5	=	Cost of pesticides per acre in rupees
X_6	=	Cost of ploughing per acre in rupees
β_0	=	Intercept
U	=	Error term
β_1 to β_6	=	Partial regression coefficients

Joan Robinson model of economic growth has been used to estimate profit rate between contact and non-contact farmers. That is—

$$P = \frac{\mu - \varpi}{\theta}$$

where,

θ	=	Capital labour ratio
ω	=	Real wage rate
P	=	Profit rate
μ	=	Productivity of labour

To identify the factors associated with variation in yield between contact and no-contact farmers cultivating ADT-36 variety of paddy, linear regression model has been used.

Table 17.1: Determinants of Yield (Paddy) among Contact and Non-Contact Farmers

Sl. No.	*Variables*	*Contact farmers*	*Non-contact farmers*
1.	Intercept	1104.64	3593.62
2.	Farm size	15.37-(2.90)*	42.78(3.21)*
3.	Seed cost	-13.26(-3.61)*	-37.39(14.41)*
4.	Labour cost	0.46(3.50)*	-0.03(-0.37)
5.	Fertilizer cost	0.24(2.05)*	-0.02(-0.23)
6.	Pesticides cost	3.42(5.89)*	-0.01(0.09)
7.	Ploughing cost	0.10(0.75)	0.05(0.65)
8.	R^2	0.94	0.69
9.	'F' test	194.63	126.21

*Significant at 5 per cent level (figures in parentheses represent t-values).
Source : Statistical analysis of data collected from the study area.

Table 17.1 reveals that, all the independent variables except seed cost have been found positively related to yield. They jointly explain about 94 per cent of the variation in yield per acre for the contact farmers cultivating ADT.36 variety of paddy. variables of farm size, labour cost, and cost of fertilizer and pesticides emerged statistically significant at 0.05 per cent level of significance. It explains that by one unit increase in these variables could increase the yield by 15.37, 0.46, 0.24, 3.42 kg per acre respectively. As the estimated F value is more than its table value at 0.05 per cent level, the overall regression model is statistically significant. The input seed is negatively related with yield which indicates that there is over usage of seeds per acre. This is true that the Department of Agriculture advised that the seed requirement is 24 kg per acre for ADT36 variety of paddy where as it is 32 kg on an average among the contact farmers.

In the case of non-contact farmers only one independent variable is found positively related to yield, namely, the farm size. Seed cost is negatively significantly related to yield. They jointly explain about 69 per cent of the variation in yield per acre for the non-contact farmers cultivating ADT36 Variety of paddy. Farm size emerges statistically significantly at 0.05 per cent level. It explains that by one unit increase in the variable, yield could be increased by 42.78 kg. The negative value of seed cost indicates that seed is used more than its requirements. The average quantity of seed used by the non-contact farmers is 50 kg which is double the quantity prescribed by the department. This indicates that seedlings may lack vigour and thus it leads to poor yield.

Table 17.2 : Capital Output-Ratio, Productivity and Profit Rate of Paddy Cultivation among Contact and Non-Contact Farmers, Values (in Rs.) per Acre

Sl. No.	*Cost of Production and Output*	*Contact farmers*	*Non-contact farmers*
1.	Yield	11244.00	8229.00
2.	Labour cost	1599.70	1883.30
3.	Fertilizer cost	1640.46	1669.00
4.	Ploughing cost	1061.60	1318.77
5.	Pesticides cost	391.56	366.83
6.	Seed cost	159.83	245.33
7.	Total (2 to 6)	4853.15	5483.23
8.	Capital output ratio	0.43	0.68
9.	Productivity	2.31	1.50
10.	Yield rate (Rice)	1405.83	1028.62
11.	Rate of Profit	1.28	0.45

Source : Computed.

Table 17.2 reveals that the yield rate (rice per acre), productivity and profit rate are greater among the contact farmers than among the non-contact farmers. The capital-output ratio of the contact farmer is less than the non-contact farmers. Lower value of capital output-ratio indicates greater productivity among the contact farmers.

Conclusion

The study reveals that farm input use efficiency, productivity and rate of profit are greater among the contact farmers than the non-contract farmers. The result implies that the farmers under the guidance of agricultural extension officer cultivate more effectively wherein the productivity and the rate of profit are also high. On the other hand, the level of productivity and the rate of profit are low and the capital output-ratio is high among the non-contact farmers. To conclude, the proper implementation of agricultural extension education and training will bring larger output and better yield to the farmers.

Agricultural Extension Education and Training must make agriculture attractive and profitable. The effective implementation of extension education system should stop the problem of rural urban migration and the conversion of agricultural land into non-agricultural activities. Today, agriculture becomes unattractive and the farmers migrate to cities. Their arrivals in the cities expose them into the inhuman urban life in the urban slums. Thus, the rural migrant lost their originality and their real life in the urban ocean.

Therefore, the agricultural extension education system should make rural people feel proud of being a farmer. The farmers must realise that the rural development is by the farmers and for the farmers. The scope of the extension education system should be widened in order to retain the rural spirit among the farmers and to ensure the availability of agricultural land for cultivation.

REFERENCES

Larkk, Eicher and Staatz (ed.), *Agricultural Development in the third world*, John Hopkins University Press, London 1984, pp. 335-347.

Mad Nagir Shamsuddin Alias Radan, Fred Liew, Productivity and Growth of Malaysian Agriculture Sector, *the Asian Economic Review*, Vol.41, No. 1, April 1999.

Schultz, T.W., *Investing in people, the economics of population quality,* Hindustan Publishing Corporation, Delhi, 1981, p. 4.

S.N. Sukumar, Agricultural Extension Education and Training on Entrepreneurial Ability of Farmers—An Intra-regional analysis of Vellore District, Unpublished Ph.D. thesis, University of Madras, Chennai, 2000.

18

ECONOMIC, TECHNOLOGICAL AND INSTITUTIONAL ISSUES OF WATERSHED MANAGEMENT—A CASE STUDY OF PAZHAYAR BASIN IN KANYAKUMARI DISTRICT

A. Meenakshisundarajan and *C.A. Sham Shankar*

Introduction

A Watershed is a geo-hydrological unit, referring to the catchment area of a drainage system. It has been accepted as a scientific unit for area development. The focus of such development will be on sustainability, envisaging a multi-dimensional approach. This indicates integrated development, which includes agriculture, horticulture, grass lands, forest lands, generation and augmentation of water resources and planned land use system. Evidently, such a holistic effort raises economic, technological and institutional issues. Some such issues are discussed in this paper, with reference to a particular watershed, viz. Pazhayar river basin of Kanyakumari district.

The river Pazhayar is as old as Tamil civilization; it has been claimed as the extant of river Pahruli of the Sangam classics. Its origin is to be traced at the southern slopes of the Mahendragiri peak. A number of small streams like Nachukal odai and Eassakkuthottam odai confluence at Chrulakode and form the origin of the river. From there, it flows southward; in the course, many small rivulets join it and at Vadasery (Nagercoil), it bends a bit towards the east and finally, it empties into the Arabian sea, at Manakudy lake. Tharuvayar, Ulakkaruviar, Alathuraiar, Koyu odai, and Poigaiar are the chief tributaries of Pazhayur. The total length of the river is 37 km; the river basin spreads over the entire 117.4 sq.km midland region of Nanjinadu (Thovalai and Agasteeswaram taluks of Kanyakumari District). Left to itself, depending on its own flows, the river can cater, naturally, to only limited areas. One Pandyan king, realising the feasibility to bring more lands under the plough in the fertile Nanjinadu, built a dam across the river Parali, that originates from another mount flowing westward, diverted the waters eastward, excavated a 2 miles canal and connected it with Pazhayar. The dam and canals are called Pandyan dam and Pandyan kal. Of course, this may be the first attempt towards linking of rivers in the south.

The pressure for the river water was continuing. So, again, the great Marthanda Varma of Travancore, built another dam—"Puthen dam"—downstream the Pandyan dam in 1747 A.D. This augmented further the flow of Pazhayar.

Still later, Sree Moolam Thirunal, the King of Travancore, constructed a dam across R. Kodayar at Pechiparai (1906) and diverted its water also through the Pandyan kal.

His successor Sree Chithirai Thirunal, built another dam upstream Pandyan dam at Perunchani (1953) and its waters were also diverted ultimately to Pazhayar.

There are two contour canals—the Thovala Channel and Ananthanar Channel—on the right and left wings of the river basin. These two, besides carrying water from the Pandyan kal, take the drainage of the adjoining hill streams as well. The water that drains from these canals flow into the low-lying Pazhayar. Similarly, the tributaries that conjoin with Pazhayar on the way, also drain their surplus into the Pazhayar. In this way the basin has emerged into a compact watershed, with the two contour canals on either side of the ghats, up above them, as the ridge lines.

Economic Issues

The availability and use of water, helps to increase cropping intensity and induces changes in cropping pattern; these two could increase agricultural production/productivity. The dynamics of agricultural development in the two taluks of Thovalai and Agasteeswaram which constitute the broader basin of the river

Pazhayar over a twelve year period (from Fasli 1400 to 1411) provides an interesting reading. The area under cultivation is depicting a declining trend in both taluks—from 14021 ha to 13029 ha. in Thovalai and from 25081 ha to 22440 ha in Agasteeswaram. The decline is larger in the case of paddy—from 9709 ha to 8145 ha. in Thovalai and from 1500 ha to 1043 ha in Agasteeswaram. In the mean time, the areas under other crops have increased from 4312 ha to 4884 ha in Thovalai and from 1007 ha to 1978 ha in Agasteeswaram in the twelve year period. Thus, there is a definite change in the cropping pattern. The reasons are worth investigating. Agriculture as an economic proposition involves a comparative study of the returns from the cultivation of competing crops. Studies have shown that net income as a percentage of total cost of cultivation is the second lowest for paddy (50 for HYV and 46 for Local variety). On the other hand, banana (105 per cent) and coconut (90 per cent) promise very high rate of net returns. These economic considerations explain the progressive decline in paddy area. Comparatively higher cost of labour in the district is another factor that prompt the farmers to switch off paddy cultivation (as the crop demands more labour, next only to sugarcane) and prefer a perennial crop like coconut. The level of wages for most agricultural operations, is the highest in the district. Again, the term loans extended by the long-term financial institutions (LDB / NABARD) lured the cultivators to shift from paddy to coconut. The Coconut Development Schemes of the LDB shall be pointed out in this connection. The paddy farmers have access to short-term loans only. But, the coconut farmers have access to medium/long-term loans for planting, as well as subsequent maintenance.

The intrusion of banana into the traditionally paddy area has led to wastage of water and misuse of land. Banana requires water throughout; even when the paddy farmer wants to keep his land dry/fallow (in summer), banana land is to be watered. Further, banana cultivation disfigures and exhausts the soil; extension of banana cultivation by leaseholder (who lure the land owners by offering hefty rents, ranging from Rs. 10,000 to Rs. 15,000 per acre) actually quickens this process.

Technological Issues

One of the technological means to make optimum use of water is to make appropriate adjustments in the cropping calendar.

The traditional cropping calendar, followed for long, in the Pazhayar basin was as follows. The first crop, raised under semi-dry condition, was sown in late April—the 10th day of Tamil month Chithirai (about April 14) was considered as the auspicious day for sowing. Utilising the summer showers in March-April-May and/or releasing waters stored in the tanks, the seeds were sown broadcast, covered by ploughing and compacted. Levelling and compacting were repeated on the third, fourth, fifth and seventh day of sowing. This compacting helps to preserve moisture for a long time and, in the meantime, checks the growth of weeds. For about a month, the crop grows with the available soil moisture. No irrigation is required during this period, and, better drying means, better yield as well. And then onwards—from late May/early June onwards the field is to be flooded with water. The local variety of paddy—*samba*, —with a maturity of 120 days, was sown during the season, which could be harvested by early September (in the Malayalam month of Kanni, and hence known as *Kannipoo*—Kar season).

The second crop was raised in August-September, and the most familiar wet transplanting cultivation was practised. The crop was harvested by mid-February (in the Malayalam month of Kumbam and hence known as Kumbapoo Pisanam season).

The time structure of this cropping pattern adjusted well with that of the monsoons. The dry period corresponded with the almost rainless month of May. The watering of the fields begins at the beginning of June, with the onset of the south-west monsoons. Harvesting was done by mid-September, before the north-east monsoon sets in. Similarly, ploughing and transplanting operations for the second crop were started when sufficient water was brought in by the north-east monsoon and, its continuation till December, assures continuous water supply for the crop. Harvesting materializes in February, which is the driest month of the year.

Thus, there was a rough correspondence between the crop and the monsoon cycles—a "synchronization of the monsoon and the crop cycles".

The operation rules of the irrigation system were also adjusted with the local cropping calendar. As seen before, artificial watering (irrigation supply) starts only after the 30th day of sowing of the first crop—that is by about 24th May. Irrigation operation rules of the erstwhile Travancore state, laid down that the dam (Pechiparai) will be opened on the 5th of Tamil month Vaikasi (at about 20th May). It will take five days for the water to reach the tail end.

This schedule corresponds very well with the crop water requirements of paddy. The tiller initiation requirements of Kanni crop are met with summer rain/tank water. The primordium initiation is provided for by making available two inches of water for 60 days (between the 30th and 90th days) and the higher requirements of the flowering and grain forming stages were met with three inches of water for the next 20 days. The harvesting was completed before the onset of north-east monsoon. Similarly, the north-east monsoon rainfall, along with the release of an inch of water for 90 days after transplantation, and two inches for the next 30 days, meet the water requirements of Kumbam crop. The harvest could be completed in the no-rain period of January-February.

This alternation of semi-dry and wet-transplanting system of cultivation helps to avoid water logging in the fields and keep up the conditions of the soil. So, the cultivators prefer it. That semi-dry technique economises the use of water is supported by basic agronomy. Of the water requirements of paddy, 20 per cent is needed at the nursery stage; another 21 percentage at the transplanting stage, 22 per cent at the post-transplanting primordian stage, 14 per cent upto the grain forming stage and the rest (23 per cent) till it fructifies. The semi-dry technique absolves the need for water in the first two stages (20 + 21 = 41 per cent of the total need). Understanding these merits of semi-dry cultivation, in times of inadequate summer rains in April, the villagers used to empty the tanks, using the stored-up water to get the soil moistured and fit for dry sowing.

Again, the wet transplanting system practiced in the Kumbam-poo also helps in economizing the use of water. Raising nurseries over limited areas (such as in a corner of the field) requires only little water (compared with the alternative of wet sowing which requires more water to be spread over the entire field). However, wet transplanting puts up labour costs.

However, these operational rules have undergone some radical changes, consequent upon the transfer of Kanyakumari district to Tamil Nadu in 1956, with the reorganization of States on a linguistic basis. For some years, the old rules were followed. However, in 1977 the Tamil Nadu Government, while dealing with the question of delegating power to PWD officials for declaring opening up of reservoirs for irrigation purposes, grouped the irrigation systems of the state into three categories, with reference to their size and importance. The systems under category I shall be opened only with the prior order of the government. Kodayar system was put under this category. The normal date for opening up of the reservoir for irrigation was put at first of June. It was also laid down that whenever the seasonal conditions warrant, a postponement of it could be done, by giving due intimation to the ryots in advance.

The order of the government aiming at uniform rules throughout the state was only a source of hardship to the ryots of Kanyakumari district and of friction between them and the department. It was reported that "after the formation of the district in 1956, the new officers posted to the district dealt with problems without knowing the peculiar features of the district. They treated the irrigation system of Kanyakumari district like those in other parts of Tamil Nadu". Actually, the new order put the traditional cropping calendar out of gear. The dam is scheduled to be opened (instead of the customary 20th May) on first of June. The delay denies timely watering of standing crops used to be sown in late April. When the tanks were also emptied for getting the lands sown, this will create very serious problems.

The officers suggested an alternative cropping cycle—the first season operation starts from first June (the date of opening of the reservoir) and to end in October, and the second, from the last week of October to end in February. Wet cultivation is to be practised in both the seasons.

Two important drawbacks of this arrangement are evident: (1) North-east monsoon rains in October seriously affect harvesting, resulting in heavy losses. (2) The arrangement is based on the assumption of a particular date of opening of reservoirs, which could be postponed.

The "encroachment culture which had permeated our society in the state" is yet another problem to be faced by the irrigation managers. It should be pointed out that the root cause for encroachments lies in the mismanagement of tanks, channels and rivers. Allowing the tanks and river beds to silt and permitting the overgrowth of water plants over such areas, not only arrests the flow of water in the water course, but also, besides restricting its width and depth, tempts the land hungry to make profitable use of such "no man's" land.

Encroachment of river banks reduces the river discharging capacity even in normal years, flood water runs into the channels at the time of rainy season, causing flood damages to the Pazhayar river system, as well as to the crops, and heavy loss to the public property.

The need for evicting the encroachments and desilting the tanks has another dimension as well. It is reported that, in the district, every year 10026 M cft. (in Pazhayar alone 4256 M cft.) rain water is wasted into the sea. This rain water harvested and stored up in the desilted tanks will help to stabilize the cultivation operations in the district—semi-dry cultivation of first crop strengthened, and water stress at second crop final stages avoided.

Institutional Issues

The two institutions that go with watershed management are the state and the water users, with their laws and conventions, functionaries and organizations.

A review of the development of Pazhayar water resources reveals an important trend. Each and every effort towards augmentation of water flows in the river has been followed by a diversion of water to some new areas—Puthen dam followed by Ananthanar and N.P. channel, Petchiparai followed by Thovalai channel and Radhapuram Channel. Such draining away of the fragile surplus to one water course from another is not helpful for any optimum utilisation of water.

The introduction of State level irrigation rules to the Pazhayar basin (as seen before) has distorted the traditional crop calendar (which was admirably synchronised with local rainfall calendar), resulting in economic losses. As such, better water management presupposes decentralised irrigation rules which are relevant to the local situation.

Turning to the role of the water users in watershed management in the Pazhayar basin, in most villages there are the *Puravus*, the main functions of which are filling of tanks and opening of the sluices of tanks. In times of scarcity, the *puravu* arranges for equal sharing of water by the people. This is executed by *Kandottus* appointed specially for the purpose. Turning to the organizational pattern of these bodies, most of them are just informal: all villagers participate in the deliberations of the *puravu*. It is just a community body.

A SWOT analysis of the functioning of the *puravus* in the area has some interesting findings:

1. The office bearers of the *puravus* are elderly (whose voice is respected), and impartial in making decisions. This makes the settlement of conflicts easier and effective.
2. The absence of any membership fees/annual contributions by the *ayacutdars*, make them feel that they get some services done freely, without any costs.
3. As they get substantial amount from duck tending, the *puravus* somehow find both ends meet.
4. In recent times, they voluntarily move with the times, accommodating the neo-farmers at the helm; there are *puravus* where the traditional land owning class has given way to the socially weaker classes.
5. The *puravu* heads value the most, the opportunity to act as a guardian of natural resources and the protector of the environment; and these enhance their strength.

The following weaknesses of the *puravus* are also found:

1. Only two (of the eight) *puravus* in the Pazhayar basin are registered ones. All others are informal; they are simple community organizations without any rules and regulations to follow.
2. The *puravu* office bearers are the most scared of lack of income to finance their functions. They fear that, in the near future, their income from duck tending will go down with the expansion of coconut cultivation and tractor tillage in their area.
3. The problem of accountability is the other factor that threatens the sustainability of the *puravus*.
4. The *puravus* are basically community bodies. They could function effectively only so long as the community is a homogenous lot.

The need for peoples' participation in watershed management was felt by the state and the World Bank, while offering funds for irrigation projects, stipulated the finding of Water-users Associations at the local level, as a precondition for such aid. The Tamil Nadu Farmer's Management of Irrigation System Act was passed in 2000 to facilitate the finding of such WUAs. The purpose of this Act is "to promote and secure distribution of water among the users, adequate maintenance of the irrigation systems, efficient and economical utilization of water to optimize agricultural production by involving the farmers, and inculcating a sense of ownership of irrigation systems in them, in accordance with the water budget and the operational plan". The Act provides for compulsory membership of farmers in an irrigation system: under the Rules of the Act, the WUAs are given a managerial subsidy (of Rs. 450 per ha. as against the Rs. 50 per ha contributed by the members). In organizing these WUAs, the work of NGOs is also solicited.

In the Pazhayar basin, 8 such state-sponsored WUAs (sangams) are founded. An intensive study of the working of these WUAs reveals certain truths. All the Sangams are only in paper; Not even a single sangam has contributed its share to get the managerial subsidy, provided by the state. These are, in fact, only '*avatars*' of erstwhile *puravus*. It seems that the traditional *puravus*, not insisting on any membership subscriptions, run by elderly and honest men and adjusting themselves to social change, cannot be easily dislodged.

It is also found that the State-sponsored associations can click only if the water users really and effectively participate in its activities. This presuppose the awareness and involvement of water users. No government GO can bring up a full-grown people's organization. The government has invited a NGO (Thiruvalluvar Social Service Society) to organize *sabhas* in the district. Excepting the sangam heads, others have no knowledge of this particular NGO and its work. Absence of this awareness and involvement is seen reflected in the sangams remaining moribund.

These sangams have to face the following threats. (1) Once the funds and powers of these sangams are augmented, it will attract the entry of politics and politicians into the affairs of watershed management, which will spell doom to the smooth functioning of the sangams. (2) Sangams are sponsored with the ultimate aim of transferring the managerial responsibilities of water bodies to the water user. This means absolving the powers of the PWD, and the bureaucracy, as it is, is not prepared for such a dethronement. (3) The problem of accountability, lack of coordination and local caste and family dissidence may also make the sangams unsustainable.

Actually, watershed management is something that is more than irrigation management. Domestic consumption has been regarded as the first claimant of water. It is the Pazhayar that caters to the drinking water requirements of all the villages on its way. Deep wells are excavated on the river bed and water is pumped to big overhead tanks, from where it is distributed to the villages. At the Chattuputur dam site itself, there are five such overhead storages, and excavation is presently going on for another nine at Veeranamangalam dam site. It is feared that the priority given to the supply of drinking water to the towns and the resultant mining of ground water in neighbouring river/tank beds, in course of time, will lead to total depletion of potential aquifers, which will reflect on the total availability of water for agriculture.

An allied problem is this: The PWD leases the right to rear

fishes and lotus flowers in the local tanks, to private parties. They never mind to preserve the quality of water. Indiscriminate use of fish feeds by the fish rearers and dead lotus leaves and flowers, spoil the water. Further, these leaseholders want to keep water in the tanks throughout the year, while the cultivators require to empty the tanks in summer, in order to prepare for dry farming in the next season. This invites clash of interests. There is yet another thing as well. Tank water has become so contaminated, that the villagers cannot make use of it for bathing. So, they were forced to take piped drinking water for bathing as well. This means a rise in the demand for water and a consequent increase in pumping charges to the village panchayats, who are finding it beyond their means to pay even their bill for street lights!

Integrated watershed development includes generation and augmentation of water as well. Kanayakumari district as a whole, and the Pazhayar basin, in particular, is blessed with both south-west and north-east monsoon rains, thanks to the lush forests at Western Ghats. However, the aged forests were axed and such denuded areas were transformed into money spinning rubber plantations—the biggest among them is under the State sector! It is complained that such thoughtless pillage, resulting in the loss of lush forests has seriously affected the quantum of rainfall in the district—There is a clear decline in the average rainfall in the district —1720 mm in 1974 and only 1456.8 mm in 2000. Further, in recent years, the actual quantum of rainfall seldom reaches the average level. Over a period of 16 years (1985-2000) the actual (1629 mm) exceeded the average, only in one year (1991); in all other years it was less than the average. While seasoned public opinion attribute the ruinous selfish exploitation of forest lands to this deplorable state of affairs, the planters' lobby says that only 'depressions in the bay' (and not lush forests) cause rainfall!

---

* Based on a Ph.D. dissertation entitled, *Development and Management of Water of Pazharar river basin (Kanyakumari District)—A Study of the Technological Economic and Institutional issues,* by CA. Sham Shankar submitted to Manonmaniam Sundaranar University.

19

RAINWATER HARVESTING : A NEED OF THE HOUR

M. Soundarapandian and *N. Prabha*

Introduction

Rainwater harvesting is the process of collecting, storing and using rainwater as a primary or supplementary water source. Here it refers to the small-scale collection of rainwater from roofs primarily for use as a domestic, potable water source.

As water source, the major advantages of rainwater collection systems are: In most areas, the quality of rainwater is excellent; the concept is simple and the system can be easily built and maintained and the ability to operate independently of outside systems renders this technology viable even in remote areas and difficult terrains.

Rainwater harvesting in brief is the deliberate gathering, concentration, and storage of water running off surface on which rain has fallen directly. The RWH can be applied in many areas but has particular importance to arid and semi-arid regions and remote isolated habitations in difficult terrains where it may often provide the only feasible solution for an improved water supply.

Benefits of Rainwater Harvesting

- Water harvesting helps smooth out variation in water availability by collecting the rain and storing it more efficiently in closed stores or in sandy riverbeds.
- Water harvesting provides an alternative source for good quality water seasonally or even the year around.

- Privacy and a feeling of 'home production' given more satisfaction for RWH.
- It also avoids conflicts associated with shared facilities among neighbours.
- It saves children's time who generally go for water fetching and result in minimization of school drop-out rating.
- This would also help to increase family income labour in total is getting saved.

Rainwater Distribution

Sea and Ocean	-	97.2%
Water in glaciers and ice caps	-	2.14%
Underground water	-	0.16%
Surface water	-	0.009%
Soil moisture	-	0.005%

It clearly indicates that the rainwater whatever received is lost in many ways. In such a situation there is no other option except finding measures to harvest rainwater. Depending on the local environment, water harvesting would provide a supplementary supply, an alternative supply or only feasible improved supply. This would help especially the rural areas falling within arid and semi-arid zones. This strategy of water harvesting is based on the principle of least interference, which is necessary for achieving the said objectives. Because of mutual benefit, the villagers through co-operation can institutionalize the water distribution, utilization and water shares.

Structure of Rainwater Harvesting

The structures required for Harvesting Rainwater are simple, economical and eco-friendly. In coastal areas over extraction of groundwater leads to saline water intrusion. Groundwater in many areas is either of poor quality or not is potable due to the presence of clay or anthropogenic pollutants. Recharge of groundwater using rainwater helps to improve the groundwater quality considerably.

In a normal year, India gets one per cent of the total global rainfall on 2 per cent of the geographical area. Much of the water is wasted for want of adequate schemes for conservation and proper storage. The annual per capita water availability in the country is estimated to be varying in the range of 300 to 13,754 cubic metres. The national average was 1.829 cubic metres in 2001 and if likely to decline to about 1.557 cubic metres by 2015 due to increase in population.

Assessing the Viability of Rainwater Harvesting

1. Study the need for procuring rainwater low water table, high mineral content in the subsoil water and satisfactory rainfall are the factors that favour the technological option to harvest rainwater.
2. Study the rainfall rate and its distribution over a period of five years. The annual rainfall, the peak periods of rainfall and the dry-spell periods need to be studied carefully to assess the economic viability of the technology.
3. Study the availability of spaces from which the rainwater can be conveniently harvested and the specific features of the same so as to decide on the instaurations needed, for the systems.
4. Estimate the available roof area. The length of the building multiplied by its width will give the total roof area.
5. Estimate the quantity of water procurable from the roof surface by applying the following formula.

Rainwater procurable = Rainfall in mm × roof surface in m^2 0.8 litres (where 0.8 is a constant derived from consideration of losses through evaporation, leakage act, during water procurement).

Parts of a Rainwater Harvesting System

Catchments

A well-maintained roof of any shape forms the catchment's area for rainwater collection terraced and titled roofing proves to be highly satisfactory. Thatched roof adds a slight colour to rainwater collected from it but does not affect its chemical quality.

Valley Gutters

The valley gutters are made of G.I. sheets of 18 or 20 gauge bent in the shape of a semicircle of 15-20 cm (6 to 8) diameter and length equal to that of the eaves of the building. Readymade gutters made of asbestos—cement or burnt clay can also be used. M.S. rods are fixed to alternate rafters, using channel clips; in such a way that their bent end protrudes beyond the eaves the valley gutters are seated on this bent portion. This work has to be done by a skilled worker. The gutters showed slope gently towards the outlet point to facilitate easy and quick discharge of rainwater to the destination without spillover losses.

Conduit-Pipe

The conduit pipe can be of AC/PVC/stoneware of 10-15 cm diameter. The size of this pipe depends on the amount of water to be carried down at the peak period of rainfall.

Foul-Flush Separator

The first run-off from roof surface contains impurities like dust, bird-droppings, decayed organic materials etc. so it must be discarded to ensure the purity of the rainwater stock at the collection point. The mechanism that is designed for this is referred to as Foul-Flush separator.

Role of Panchayat and People's Response Towards RWH

Panchayat have been the backbone of the Indian villages. Panchayats are the first institutions of local self-governance where democratic process begins. Panchayat can play a vital role in implementing RWH in villages, through creating awareness among the people's regarding RWH. Rainwater Harvesing promised to be an extremely beneficial method to avoid water scarcity in villages. Thus the decision-makers in Panchayat can give the necessary impetus for kick-starting the campaign to make it a people's movement. Accordingly, people participation in implementing the rainwater harvesting are to be encouraged by some institution like GOs and NGOs. Response of people towards RWHs are:

- It should be constructed across the river in order that its storage could be tapped through pumps for irrigation purposes, to improve their economy.
- A part of the storage may be utilized for human and cattle.
- The dam would exclusively be utilised for providing water to humans and cattle.
- The height of the structure should be such that it should not submerge land and property.
- After the construction of the structure there should not be any damage to and during high flood.
- Technology should be simple, sound and labour intensive. So that it can benefit local population.
- The top of the dam, if possible, should act as a cause away.

In short, the society was demanding economic rationale, social justification, long-term sustainability and environmental safeguards for wellbeing of its inhabitants.

How much Water can be Harvested?

Amount of rainwater harvesting depends on the annual rainfall, total area of collection and the amount of recharge into the ground.

Average annual rainfall of Chennai	- 1200 mm (1.2m)
Amount of rainwater collected in	- 223× 1.2=267cv. m/ year
One ground plot (2235 q.m) per year	- 2,67,000. litre/year (approx)
Volume of water recharged into the Ground (considering of 60% effective Recharge) (approx)	- 1,60,000 litres/year

This means that 1,60,000 litres/year is replenished out of the total quantity of water extracting from the ground.

Rainwater Harvesting Indian Experience

In the Indian context the term "Rainwater Harvesting" connotes collection and storage of rainwater and also other activities aimed at harvesting surface water and groundwater, prevention of losses through exploration and seepage, and all other hydrological studies and engineering interventions aimed at conservation and efficient utilization of limited water endowment, of a hypsographic unit such as a watershed or a geomorphic basin.

The water harvesting can be effected by :

- In-sit-harvesting: nadi, tank, sandfilled reservoirs, ponds, roof top collections, Hill top collection, plat form etc.
- Storage of water in aquifers (natural underground reserves) percolation tanks, khadins, anicut/check dams; sub-surface dams, barriers, injection wells.
- Soil conservation methods which help in increased groundwater recharge also : Gully plugging, contour bunding, afforestation, contour trenching, land levelling, and bunding of fields.
- Enhancement of surface run off collection: catchments treatment.
- Evaporation control: Chemical films, hydrophobic coating material.

Direct storage of drinking water is necessary in the areas. Where rainfall is scanty and the water bearing geological formation is not suitable. Groundwater storage is the best method of Water Harvesting as it does not involves filtration of surface water but also safe from evaporation losses and natural catastrophes.

The land development, soil conservation methods and afforestation methods whose main function is to improve the ecological and environmental conditions. Indirectly, improves the groundwater recharge and increase the surface water storage.

Summary

Water is one of the natural resources without which the existence of human being is impossible. Almost all-sectoral activities, including economic sectors like households, agriculture, industry etc., depend of the availability of water. Once water was considered as a free good, but in recent years, water is treated as a priced good. The demand for water in different economic sectors has been continuously increasing over supply of water, resulted in water scarcity and water crisis. The problem water scarcity and water crisis have necessitated in conserving water and managing water bodies, for which different institutions have been effectively involved in managing the water resources. One such effort is Rain-water Harvesting System, through which water can be saved, conserved and stored in more scientific way.

REFERENCES

Kausal Kishore, "Rainwater Harvesting", *Kissan World.* Vo1.31, No.2

Khan, M.A. "Rainwater Harvesting Technology for Sustainable Water Management", *Kurukshetra.* Vol. 49, No. 8, May 2001.

Krishnamurthy, K., "Rain Water from Western Ghats to Mitigate Droughts", *Kissan World,* Vol.29 No.2 Feb. 2002.

Arunadevi "Rain Water Harvesting Measures in Urban Region", *Kissan World.* Vol.30, No. 11, 2003.

Gupta, V.S. "Water Conservaton through Watershed Development", *Kurukshetra.* October 2003.

Raj Pravin T., Maheswari M., "Water Harvesting", *Kissan World.* Vo1.28 No.6. June 2001.

20

A STUDY ON GREEN PRODUCTIVITY IN AGRICULTURAL SECTOR AND ITS IMPACT

V.S. Ganesamurthy and *V. Nachimuthu*

The new agricultural strategy consisting of Intensive Agricultural District Programme (IADP) and High Yielding Varieties Programme (HYVP) necessitated heavy investment in Seeds, Fertilizers, Pesticides and Water. This strategy has been called by various names: Modern Agricultural Technology (MAT), Seed Fertilizer (SF), Water Technology (WT) or simply Green Revolution (GR).[1] Though the Green Revolution effectively banished food shortages, further yield increment of high yielding varieties became difficult and smaller. The productivity levels soon flattened due to inappropriate use of water, fertilizers, and plant potation chemicals (pesticides) led to deterioration of soil health, ecology and environment. Thus the faster rate of growth of agriculture led to the realization that technology is crucial for providing answers to the many challenges facing Indian Agriculture[2]. For example, the gross irrigated area has increased from less than one million hectares per annum before green revolution (mid-60s) to about 2.5 million hectares per annum during the 1970s. But the scope for further increase has been limited since then due to shortage of arable land. The total food grains (cereals and pulses) production has increased from 48.1 million tonnes in 1950-51 to 211 million tonnes in 2001-2002. The remarkable progress, which the nation has achieved in food self-sufficiency. While the overall growth in food grains production assisted by the technological breakthrough has been impressive. Although production has increased enormously, Indian agriculture is still plagued by low productivity.[3] As a result an issue that has come to occupy a prominent place in development on the environment. It is the time that countries put a price on the environmental degradation that economic development causes. India has been paying an enormous price for its economic growth in the form of ecological devastation. It has been estimated that in 1992 the environmental damage caused in India has been between $10 billion and $12 billion.

Methodology

Ten Years Time Series data have been collected from *Reserve Bank of India (RBI) Bulletin*. The period covers 1991-92 to 2000-2001. The following variables have been selected for the study—

1. Gross Sown Area (GSA),
2. Gross Irrigated Area (GIS),
3. Consumption of Fertilizers, and
4. Consumption of Pesticides.

Since the agricultural production is directly related to these important inputs further it is also affecting the environment particularly Soil and Water. Increased production by means of using more these inputs will directly lead to environmental degradation. The following statistical tools are used for the analysis, viz., Mean (X), Standard Deviation (SD), Correlation (R), Regression (R Square), Analysis of Variance (ANOVA) and Co-efficient.

Data Analysis

Table 20.1 shows that the pattern of land use for agricultural production during the post globalisation period between 1991-92 and 2000-01. The total agricultural production (both food grains and commercial crops) has increased from 186.98 million tonnes to 223.82 million tonnes in the early Nineties. Then it has declined to 213.58 million tonnes in the immediate proceeding year and increased to 229.27 million tonnes. The study has been found from its analysis that the agricultural total production has fully depended upon the independent variables, namely Gross Sown Area and Gross Irrigated Area. It can be proved by the simple correlation, the value is .931 and .848 for gross sown area and gross irrigated area respectively. However, it exhibits a positive and high relationship

Table 20.1 : Pattern of Land use for Agricultural Production

Sl.No.	*Years*	*Gross Sown Area (in million*	*Gross Irrigated Area hectares)*	*Total Production (in million tonnes)*
1.	1991-92	182.24	65.68	186.98
2.	1992-93	185.70	66.76	199.59
3.	1993-94	186.58	68.25	205.76
4.	1994-95	188.05	70.65	212.84
5.	1995-96	187.47	71.35	202.53
6.	1996-97	189.59	73.25	223.82
7.	1997-98	190.76	72.78	213.58
8.	1998-99	191.26	72.89	228.36
9.	1999-2000	191.87	71.98	229.74
10.	2000-2001	190.32	70.25	214.27
	Mean	188.3840	70.3840	211.7470
	SD.	2.9893	2.6587	13.4924
	R	.931	.848	1.000

Source: RBI Bulletin 2002.

between the response (dependent) and cause (independent) variables. The Mean and Standard Deviation value has been also calculated to show the average value and the extent of variation for the selected variables.

Table 20.2 reveals an interesting features of agricultural production, which has been influenced by selected agricultural inputs like consumption of fertilizers and pesticides during the period of one decade. Whereas the consumption of fertilizers has been increased from 121.53 lakh tonnes in the year 1992-93 to 180.70 lakh tonnes in the year 1999-2000. Farmers have gradually increased it to produce more agricultural production. But the last

Table 20.2 : Selected Inputs for Agricultural Production

Sl.No.	*Years*	*Consumption of fertilizers (Lakh Tonnes)*	*Consumption of Pesticide (000' Tonnes)*	*Total Production (in million Tonnes)*
1.	1991-92	127.28	72.13	186.98
2.	1992-93	121.53	70.79	199.59
3.	1993-94	123.66	63.65	205.76
4.	1994-95	135.64	61.36	212.84
5.	1995-96	138.77	61.26	202.53
6.	1996-97	143.08	56.11	223.82
7.	1997-98	167.88	52.24	213.58
8.	1998-99	167.98	49.16	228.36
9.	1999-2000	180.70	43.38	229.74
10.	2000-2001	172.78	46.26	214.27
	Mean	147.3300	57.6250	211.7470
	SD	21.7442	9.8874	13.4924
	R	.758	-0.864	1.000

Source : RBI Bulletin 2002.

year of the study has witnessed to decline the fertilizer usage because of degradation of environment due to these inputs. On the other hand, the use of pesticide has been declined from 72.13 thousand tonnes to 43.38 thousand tonnes during the study period. It may be due to the deterioration of soil health, ecology and environment. And also the agricultural department has been suggested to use organic manure and vermin compost for the protection of environment.

The study also inferred that there has been inverse relationship between the selected inputs. However, when the consumption of fertilizers has been increased, the consumption of pesticides has been declined during the study period. From the analysis, one can observed that the consumption of pesticides would not influence the production but it has affected the green productivity in the form of ecological devastation. The calculated R-value is .758 for fertilizers whereas the R-value is -.864 for pesticides. Therefore, there is a significant and positive relationship between fertilizers and production but the negative relationship has showed in the case of pesticides and production during the study period. The Mean and Standard Deviation value also calculated to show the fluctuating trends in the production, is influenced by selected agricultural inputs.

Tables 20.3, 20.4 and 20.5 are summarising the determinants of agricultural production with the help of regression analysis.

Table 20.3 : Regression Result Determinants of Agricultural Production.

Regression Equation of Y on X_1, X_2
Where Y = Total Production
X_1 = Gross Sown Area
X_2 = Gross Irrigated Area

Model Summary

R	*R Square*	*Adjusted R Sqare*	*Level of Sign*
0.933	0.870	0.833	*

Analysis of Variance (ANOVA) :

	Sum of Squares	*DF*	*Mean Square*	*F*	*Sig.*
Regression	1425.068	2	712.534	23.379	*
Residual	213.344	7	30.478		
Total	1638.411	9			

Note : *1% ,**5%, NS=Not significant

Co-efficient

Variables	*B*	*T*	*Sig.*
Y	-533.952	-3.277	0.014
X_1	3.737	2.849	0.025
X_2	0.592	0.401	0.700

Source : Compiled from collected data.

Table 20.4 : Regression Result Determinant of Agricultural Production

Regression Equation of Y on X_1, X_2
Where Y= Total Production
X_1= Consumption of Fertilizers
X_2= Consumption of Pesticides

Model Summary

R	*R Square*	*Adjusted R Square*	*Level of Significant*
0.901	0.811	0.757	*

Analysis of Variance (ANOVA) :

	Sum of Squares	*DF*	*Mean Square*	*F*	*Sign.*
Regression	1328.783	2	664.391	15.020	*
Residual	309.629	7	44.233		
Total	1638.411	9			

Note: *1% , **5%, NS=Not significant

Co-efficients

Variables	*B*	*T*	*Sign.*
Y	430.256	4.373	0.003
X_1	-0.560	-2.961	0.021
X_2	-2.361	-1.544	0.166

Source : Compiled from collected data.

Table 20.5 : Regression Result Determinants of Agricultural Production.

Regression Equation of Y on X_1, X_2, X_3, X_4
Where Y= Total Production
X_1= Gross Sown Area
X_2= Gross Irrigated Area
X_3= Consumption of Fertilizers
X_4= Consumption of Pesticide

Model Summary

R	*R Square*	*Adjusted R Square*	*Level of Significant*
0.938	0.879	0.782	*

Analysis of Variance (ANOVA) :

	Sum of Squares	*DF*	*Mean Square*	*F*	*Sign.*
Regression	1440.155	4	360.039	9.080	NS
Residual	198.257	5	39.651		
Total	1638.411	9			

Note : * 1 % , **5%, NS= Not significant

Co-efficient

Variables	*B*	*T*	*Sign.*
Y	-394.233	-0.654	0.542
X_1	3.428	1.098	0.322
X_2	0.437	0.243	0.818
X_3	-0.239	-0.607	0.570
X_4	-0.616	-0.654	0.660

Source : Compiled from collected data.

1. If we consider the gross sown area and gross irrigated area, they are influencing the growth of agricultural production with R^2 value of .870 at 1 per cent level of significant and ANOVA has also been statistically significant by the value of .001. Therefore, these two inputs may influence the growth of agricultural production.
2. If we consider the consumption of fertilizer and pesticides, the odd factor has been influenced whereas the even factor could not lead more production. But there is some list correlation between dependent variables. The R^2 value is .811 and these factors have also been statistically significantly at 1 per cent level. The computed value of coefficient is .021 for fertilizer may only increased production whereas the consumption of pesticides has been led to the negative relationship between the variables.
3. If we have applied these four variables to agricultural

production, the R^2 value has showed a positive correlation of 87.9 per cent of very high degree at 1 per cent level of significant with variables. But in the co-efficient, T value of fertilizer and pesticides are negative, on the other hand, T value of the gross sown area and gross Irrigated area, are positive, therefore, these four compiled variables are statistically insignificant.

Conclusion

Hence, the study has concluded that the chemical materials have increased certain level after, it has declined if we use chemicals and pesticides as well as it has led to the degradation of environment in the agricultural sector.

Suggestions

Based upon the study, the agricultural activities should adopt new strategy with use of organic manure and vermin compost fertilizers, instead of using chemical pesticides and new high yielding seeding.

REFERENCES

General Studies Manual 2004, Tata McGraw-Hills Publishing Company Limited, New Delhi. pp. F5& F33.

Purohit/Shammi/Agrawal, *A text book of Environmental Science*, Student Edition, Jodhpur, 2004. pp 248-256.

Reserve Bank of India Bulletin 2002.

Ruddar Datt & K.P.M. Sundram, *Indian Economy*, S. Chand & Company Ltd. New Delhi, 2000. pp 485-487.

R.V. Badi & N.V. Badi ; *Rural Marketing*, Himalaya Publishing House, New Delhi, 2004. pp 537.

21

GREEN GOVERNANCE AND SUSTAINABLE MANAGEMENT OF WATER RESOURCE IN INDIA

Er. A.T. Narasimhan

At less than 1000 km^3/year, world's annual use of groundwater is 1.5 per cent of renewable water resource but contributes a lion's share of water-induced human welfare. Global groundwater use however has increased manifold in the past 50 years and human race has never had to manage groundwater use on such a large scale. Sustaining the massive welfare gains groundwater development has created without training the resource is a key water challenge facing the world today. During the past century, a growing corpus of experiential knowledge has accumulated in the industrialized world on managing groundwater in various uses and contexts. A daunting global groundwater issue today is to apply this knowledge intelligently to by for the more formidable challenge that has arisen in developing regions of Asia and Africa, where groundwater irrigation has evolved into a colossal anarchy supporting billions of livelihoods but threatening the resource itself.

Global Groundwater Juggernaut

Rapid growth in groundwater use is a central aspect of the world's water story, especially since 1950. Shallow wells and muscle-driven lifting devices have been in vogue in many parts of the world for the millennia. In British India (which included today's India, Pakistan and Bangladesh), wells accounted for over 30 per cent of irrigated land even in 1903 (htp://dsal.uchicago.edu/statics/1894.excel) when only 14 per cent of cropped area was irrigated. With the rise of the tubewell and pump technology, groundwater use soared to previously unthinkable levels after 1950, in Spain, groundwater use increased from 2 km^2 during 1960-2000 before it stabilized (Martinez Cartina and Hemandez-Mora, 2003). In the US, groundwater share in irrigation has increased from 23 per cent in 1950 to 42 per cent in 2000 (http://water.usgs.gov/pubs/circ/2004/cir 1268/j) In the Indian subcontinent, groundwater use soared from around 10-20 km^2 before 1950 to 240-260 km^2 today (Shah *et al.,* 2003a). Data on groundwater use are scarce however attempted to back cast the probable trajectories of growth in groundwater use in selected countries. While in the US, Spain, Mexico and North-African countries like Morocco and Tunisia, the total groundwater use pecked during 1980s or thereabouts, in South Asia and North China plains, the upward trend begun during the 1970s is still continuing. A third wave of growth in groundwater use is likely in the making in many regions of Africa and in some South and Southeast Asian countries such as Vietnam and Sri Lanka (Mope *et al.* 2003).

Typology of Groundwater Socio-ecologies

At less than 1000 km^3 per year, global groundwater use is a quarter of total global water withdrawals but just 1.5 per cent of the world's annually renewable freshwater supplies, 8.2 per cent of annually renewable groundwater, and 0.0001 per cent of global groundwater reserves estimated to be between 7-23 million km^3. Yet its contribution to human welfare is based on intensive groundwater use, each embodying a unique pattern of interaction between socio-economic, demographic and ecological variables, and each presenting a distinctive groundwater management challenge.

Groundwater has historically supplied water in numerous human settlements, urban and rural, around the world. According to one estimate, "over half of the world's population relies on groundwater as a drinking water supply" (Coughanowr, 1994). 70 per cent of piped water supply in EU is drawn from groundwater. Management presents unique challenges since, in the process of urbanization, the population of a habitat generally grows faster than its geographic span. As a result, pressure on groundwater resources underlying the habitat increases rapidly as villages grow into towns and thence into cities. The ubiquitous response combines import of surface or groundwater from a distant source, volumetric pricing, improved water supply infrastructure and service to crowd out private urban tubewells to reduce pressure on urban groundwater.

Arid and semi-arid countries in the MENA region—Saudi Arabia, Yemen, Jordan, Oman, Behrain, UAE, Iran, Lybia, Egypt depend on either fossil or limitedly renewable groundwater. Some, such as Saudi Arabia, Jordan, Yemen and Libya experimented with intensive groundwater use in agriculture to secure food self-sufficiency. However, it is increasingly realized that the use of fossil groundwater even in large reserves such as the Nubian aquifer-needs to be managed in a planned manner using different criteria than used for managing renewable groundwater. Virtual water imports, off-form livelihoods, shifting and reduction in agricultural areas, wastewater treatment and re-use, desalination are elements of strategies used to ease pressure on fossil groundwater.

In recent decades, groundwater has become increasingly important in meeting water needs of industries and industrial agriculture in many developed countries such as Spain, US and Australia. Three key characteristics are: (a) users are normally few, large and identifiable as a result, it becomes possible to create and enforce rules, norms, rights and economic incentives to regulate use by creating a formal economy, (b) using groundwater as a factor of production, generate substantial wealth which is shared by relatively small number of resource users, and (c) as a result, these attract and support scientific and technical wherewithal for intensive management of the resource and its use.

In terms of groundwater quantity and number of people involved, by for the largest growth in groundwater use has occurred in sustaining subsistence crop and livestock farming, which are the mainstay of billions of poor people in developing agrarian economies around the world such as India, Bangladesh, Nepal, China. Out of the global annual groundwater use of 950-1000 km^3, half or more is likely accounted from the resource governance viewpoint, these represent a different ballgame altogether because: (a) they are dominated by large diffuse masses of small users who are neither registered, nor licensed, operating as they do in totally informal irrigation economies untrammelled by laws and regulations, (b) unlike of Spain, US and Australia, support large number of poor people but generate little wealth in absolute or relative terms. A groundwater user in South Asia produces a gross output of US $ 400/ha from irrigating crops; in contrast, a Spanish farmer in Andalucia region generates gross output/ha of US $ 8000/ha on average but can go up to US $ 75000 (Liamas, 2003); (c) despite these apparently low returns has served as one of the largest and most patient 'poverty reduction' programmes (Deb Roy and Shah, 2003) in recent decades and (d) since science, technology and management tend to get attracted to wealth generation more easily than to poverty reduction.

Numerous aquifers in the world are shared by two or more sovereign states; most of these are small but some like the Nubian with an estimated reserve of over 500,000 km^3 are huge (Puri and El Naser, 2003). As intensive groundwater use emerges in these aquifers, their effective governance becomes subject to a new class of problems needing unique institutional responses and mediating mechanisms, Management of shared aquifers between Israel and Palestine, between the US and Mexico, and amongst countries of the Nile basin who will share the Nubian illustrate these unique issues.

Groundwater and Poverty in Asia

Globally, growth in groundwater irrigation has had little to do with the occurrence of the resource if anything, led essentially by demand-pull, intensive development has tended to occur in arid and semi-arid regions with relatively poor groundwater endowments. Regions with abundant rainfall and recharge much of South America, South East Asia, Southern China—make little use of groundwater in agriculture. Intensive groundwater use, where extraction per km^3 of annual recharge is high, has also had little to do with the geology of regions.

South Asia and North China plains represent a variable anarchy functioning on a colossal scale. India, for instance, has been adding 0.8-1 million new tubewells every year since 1990 and there is no sign of deceleration in this trend. One in four of India's farmers have invested in irrigation wells most of the remaining buy pump irrigation service from their tubewell owing neighbours. Government of India claims 60 per cent of India's irrigated areas are served by groundwater wells, independent surveys suggest the figure may well be 75 per cent and even more if conjunctive use areas are included. Much the same is true of Pakistan, Nepal Terai, Bangladesh, and Hebei, Shandong and Henan provinces in the Yellow river basin in North China plains. Governments and donors have invested heavily in building major dams and canal irrigation projects in these regions but as of now, by for the bulk of the irrigation and livelihood benefits are delivered by groundwater wells. Over half of the total populations of India, Pakistan and Bangladesh have a livelihood stake in well irrigation. During 1970s, India discussed different strategies for irrigation command areas and for rainfed farming regions. Thanks to groundwater development, there are hardly any rainfed farming 'regions' or even villages in India, there are just rainfed and mostly groundwater irrigated plots.

Groundwater Governance Institutions, Law, Policies

This run away growth in developing countries in Asia exemplifies best how poverty works as the enemy of environment. High population pressure on agriculture has induced farmers to overwork their tiny land holdings in search of more livelihoods per unit of all that land has to offer soil nutrients, moisture and underlying groundwater. Widespread indications of groundwater depletion and deterioration, rising energy use and pumping costs, well failures, weakening drought protection suggest that the 'groundwater boom', which has done more to sustain the poor than all poverty eradication programmes, will burst, sooner or later. There are also environmental repercussions in the form of drying up of wetlands and streams, reduced lean season flows of rivers, salinity ingress in coastal areas. Groundwater quality issues too have assumed serious proportions in many parts of the world, irrigating with saline groundwater, as in the Indus basin and in Australia, have raised the specter of soil salinisation on large areas. People and policy-makers in many parts of the world but especially in South Asia and North China plains are waking up to the dangers of drinking poor quality groundwater high in arsenic or fluoride or other contamination.

Effective management of groundwater demand to match available recharge is considered central to sustaining intensive groundwater use. Community management of groundwater as a common property resource is widely espoused to South Asian policy-makers based, for example, on the experience of countries like Spain and Mexico. The issue is if such models can or should be transplanted without ascertaining their effectiveness on their home turf. Spain's 1985 Water Law mandated Water User Associations of aquifer level, but of some 1400 that were registered, Martinez Cartina and Hermandez-Mora (2003) could identify "only two which have actively managed their aquifers, financing all their activities from membership fees". One reason why these failed, as Liamas points out, was that these user associations mandated top down by law have been 'fraught with strong resistance from farmers'. Mexico likewise has been experimenting with COTAS (Technical Committee for Aquifer Management), these too are yet to begin playing effective role in aquifer management (Shah, Scott and Bucheler, 2004). Groundwater districts of US are often held out as a model in community groundwater management, however, the US experience itself is a mixed bag. Since 1949, Texas allowed the creation of Underground Water Conservation Districts (UWCDs) with discretionary power to regulate groundwater withdrawals and space wells as well as their production. However, Smith (2003:264-265) notes, "Although over 40 UWCDs have been created in Texas, they have not been effective managers of groundwater" and further that " ....creating groundwater districts is not in and of itself going to ensure sound groundwater management....."

Demand restriction has also been tried through a combination of pricing, legislative and regulatory action, licensing and permits and by specifying property rights. Direct regulation worked better in countries with a hard state, as in Iran, which imposed on effective ban on new tube wells in one-third of its central plains of Russia, which has banned the use of groundwater for irrigation to protect it for domestic uses. However, bans proved counterproductive in Mexico, which has issued 14 bans on new tubewells since 1948, however, "every announcement of an imminent ban stimulated a flurry of tubewell making activity", (Shah, Scott and Buecheler, 2002). Mexico has also tried, in early 1990s, creating tradeable private property rights in ground water by issuing 'concessions' to tubewell owners with pre-specified volumes of groundwater to be pumped every year. The idea was that once private water rights are created, users would have strong incentive in protecting the resource, especially if such rights were valuable and tradeable (Holden and Tabani, 2001). Concessions have led to registration of tubewells, useful in itself, but enforcing the groundwater quota has proved administratively impossible even though Mexico has all of 90,000 irrigation tubewells, compared to North China's 4.5 million and India's 20 million. China's water withdrawal permits system and withdrawal fees have not helped reduce agricultural withdrawal although it has helped control urban groundwater depletion somewhat. Saudi Arabia has begun controlling groundwater irrigation by paying farmers for supplying water to towns (Abderrahman, 2004 pers. Comm.).

In transposing the lessons from Mexico, Spain, Western US experiments to Asian contexts, several issues come up: (a) there is no evidence that these experiments have actually led to effective resource governance in Mexico, Spain or the US, Western US has been struggling with groundwater governance for over 50 years now and yet horror stories of groundwater abuse in the US galore (for a recent one, see, Glennon's book "*Water Follies*" reviewed by Jehl, 2002); (b) groundwater demand restriction has normally worked only when alternative supplies are arranged; thus many cities in North China have been able to crowd out private urban tubewells but only after surface water and providing it in lieu of pumping groundwater. Similarly, 50 years after it began depleting its groundwater, Arizona could control groundwater demand only by providing farmers subsidized Colarodo river water in lieu of pumping groundwater (Jacobs and Holway 2004:58). Spain's 2001 National Water Plan's response to groundwater depletion on its southeastern Mediterranean coast is importing surface water from Ebro river basin (Martinez Cortina and Hermandez-Nara, 2003). In effect, then, what has commonly worked is not demand management, but 'groundwater substitution' with imported water. The US has small number of large capacity pumping plants that produce 110 km^3 of groundwater for a wealth generating irrigation machine on which less than two per cent of Americans depend for their livelihood, India, in contrast, has around 20 million small pumps scattered over a vast countryside, each pumping an average 10000m^3 to irrigate their tiny parcels in a peasant economy that has 55-60 per cent of Indians as direct or indirect stakeholders. Here, resource management capacities are poor. Regulatory, agencies are skeletal and the numbers of tiny users to be regulated huge and scattered over a vast country since. Groundwater irrigation central to their livelihoods, farmers organize readily and often violently to oppose any effort that hits their irrigation economy. Above all, many environmental ill effects of intensive groundwater use begin to occur at low levels of groundwater development. Drying up of wetlands, reduction in summer low flows in rivers and streams, increased fluoride levels in groundwater arc examples. Reversing all these would require restoring pre-development conditions by cutting the present rate of groundwater use by 70 per cent or more in many regions. Even if possible, doing this would throw out of gear million of rural livelihoods and cause massive social unrest.

Table 21.1 : Structure of National Groundwater Economics of Selected Countries

Country	*Annual Groundwater Use (km³)*	*No. of groundwater structures (million)*	*Extraction/ structure (m³/year)*	*Population dependent on groundwater (%)*
India	185-200	20	9000-10000	55-60
Pakistan	45	0.5	90000	60-65
China	75	3.5	21500	22-25
Iran	29	0.5	58000	12-18
Mexico	29	0.07	414285	5-6
USA	110	0.2	550,000	<1-2

The Case of India

People, agencies and leaders are often lukewarm to 'groundwater demand restriction' approaches even as concerns about resource protection and sustainability are mounting. While learning intelligently from the experiences socio-ecologies needs to build their homegrown approaches that strike a balance between the need to protect the resource and support their poor people. India exemplifies this challenge in its most serious form. It is facing unsustainable groundwater use in western unconfined alluvial aquifers, very much like the North China plains, as well as in peninsular hard rock India where aquifers have little storage but precipitation is relatively better. Three large-scale responses to groundwater depletion in India have emerged in recent years in an uncoordinated manner, and each presents on element of what might be its coherent strategy of resource governance.

Energy-Irrigation Nexus

Throughout South Asia, the 'groundwater boom' was fired during the 1970s and 80s by government support to tubewells and subsidies to electricity supplied by state owned electricity utilities to farmers. The invidious energy irrigation nexus that emerged as a result and wrecked the electricity utilities and encouraged waste of groundwater are widely criticized. However, hidden in this nexus is a unique opportunity for groundwater managers to influence the working of the colossal anarchy that is India's groundwater socio-ecology. Even while subsidizing electricity, many state governments have begun restricting power supply to agriculture to cut their losses. Much International Water Management Institute (IWMI) research has shown that with intelligent management of power supply to agriculture, energy, irrigation nexus can be powerful tool for groundwater demand management. Research has also shown that after all its labours to create tradable property rights in groundwater and creating COTAS, Mexico has finally had to turn to electricity supply management to enforce its groundwater concessions (Scott, Shah and Buechler, 2003).

In Western India's unconfined alluvial aquifers, it is being increasingly realized that groundwater depletion can be countered only by importing surface water, Arizona style Jiangsu province in eastern China has implemented its own little inter-basin water transfer from Yongzee to counter groundwater depletion in the northern part. Similarly, one of the major user Gujarat has found the water of the famous Sardar Sarovar Project (SSP) on Narmada river is to recharge the depleted aquifers of north Gujarat and Kachchh. A key consideration behind India's proposed mega scheme to link its northern rivers with peninsular rivers too is to counter groundwater depletion in western and southern India.

In many parts of hard rock India, groundwater depletion has invoked wildfire community based mass movement for rainwater harvesting and recharge, which interestingly has failed to take off in unconfined alluvial aquifers. It is difficult to assess the social value of this movement partly because formal hydrology and popular hydrology have failed to find a meeting ground. Scientists want check dams sited near recharge zones; villagers want them close to their wells after filtering. Scientists recommend recharge tubewells to counter the silk layer impeding recharge, farmers just direct floodwaters into their wells after filtering. Scientists worry about upstream-downstream externalities farmers say everyone lives downstream. Scientists say the hard rock aquifers have too little storage to justify the prolific growth in recharge structure, people say a recharge structure is worthwhile if their wells provide even 1000m^3 of lifesaving irrigation/ha in times of delayed rain. Hydrologists keep writing the obituary of the recharge movement; but the movement has spread from eastern Ra;asthan to Gujarat, thence to Madhya Pradesh and Andhra Pradesh. Protagonists think that with better planning of recharge structures and larger coverage; decentralized recharge movement can be a major response to India's groundwater depletion because it can ensure that water tables in pockets of intensive use rebound dose to pre-development levels of the end of the monsoon season every year they have a good monsoon, which is at least twice in five years. They surmise that this is not impossible because even today, India's total groundwater extraction is barely five per cent of its precipitation.

Indian governments at Centre and State levels have been trying for decades to secure people's participation in improving the management of canal systems, water supply and sanitation systems, drainage systems and so on, but to little avail. As a result, under remote, bureaucratic management, public water infrastructure and services have steadily deteriorated. The groundwater economy, in contrast, has never suffered for want of people's participation. What it has lacked is appropriate and intelligent participation from public agencies, science institutions and the international community. Indian engineers take pride in having built some of the finest dams in the world; but India is yet to see large scale initiatives in Aquifer Storage and Recovery (ASR) as in New South Wales, or learn to operate major groundwater banking operations as in Arizona, or master the art of depleting and refilling aquifers on an annual basis as the French do with the Montpiller aquifer.

Considered from this perspective, one can stand India's groundwater problem on its head and argue that the emergence of intensive groundwater use in regions with 1000-1400 mm normal rainfall may well be a great hidden opportunity. Through their 20 million tubewells, India's farmers have created a 185-200 km^3 reservoir in the form of dewatered aquifers, which can regularly collect, store and deliver of the users 'doorstep a relatively high quality water service that in some ways is self-regulating and self-financing'. Like all surface reservoirs, the underground reservoir has limitations, but this is precisely why science and management are required. Using this opportunity would require investing in creating scientific capability and infrastructure for groundwater recharge a top priority.

Such as India and Bangladesh with significant renewable water resources. A hundred years ago, when India did not use much groundwater and the tubewell-pump-recharge technologies were not available, it was understandable for the colonial government to concentrate resources on building great canal irrigation systems. But today when wells, pumps and recharge structures are the dominant choice of millions of India's small holders, within and outside canal commands, a smart water policy might focus on devoting resources to supporting this people's rather than throwing good money, as India is intent on doing, in pursuing an irrigation development strategy based on canal irrigation that has left a great deal to be desired.

If the world's water crisis is "mainly a crisis of governance" (GWP, 2000), groundwater represents the grimmest side of this crisis in Asia. The Australian Groundwater School of Adelaide is apt in its credo, which says, "Groundwater will be the enduring gauge of this generation's intelligence in water and land management". In exploring the nature of the global groundwater challenge, this paper has:

(a) Highlighted the tremendous contribution groundwater has made to human welfare globally;
(b) Analyzed socio-ecological implications of runaway growth of groundwater irrigation, especially in some Asian countries; and
(c) Argued why groundwater governance strategies must be context specific to be effective.

Where protecting the resource is often in direct and immediate conflict with livelihood support to rural poor-presents the most complex resource governance challenge facing the world's water professionals. Groundwater managers need to learn intelligently from approaches tried in which have been evolving refined structures of groundwater governance through demand and supply side management. Their challenge, however, is to fit these approaches into the unique contextual realities.

The rural development depends completely on the optimum use of natural resources including the water resources, which are more or less the same for decades together. But the optimum use of these resources is only very limited in rural areas, because science and technology has not created a favourable climate there. On the other hand developments in science and technology result in high demand and wastage of important natural resources like fresh water. Now-a-day, we are speaking about conservation and efficient management of water, since the per capita availability of fresh water in India has come down from 5150 cubic metres in 1947 to 1700 cubic metres in 1997. India is now not at water stress (below 1500 cubic metres) but will face scarcity of water around 2025. Further, developmental activities pollute the limited available water bodies and thereby reduce the quantity of quality water available for human use. That is, technological changes neither increase productivity of water both in agriculture and industry, nor increase the quantity of good quality water available for our use. But they only help the people by making the water drawing techniques easier than before.

The actual quantity of water used in agriculture and other industries including agro-based industries is much higher than the actual requirement of the water used by other developed countries. It is because of lack of transfer of technology or the failure of suitably blending traditional technology with modern technology. For example in India 15000 litres of water is used for producing 1 kg of paddy while the actual requirement is only 4000 litres. In the same way 500 litres are used to produce 1 kg of wheat. Further, the break up for other products is: one kilogram of paper 900 litres, 1 kg. of artificial fibre—750 litres, and 1 kg of steel—750 litres of water. This is in sharp contrast to utilization of water in China where only 230 to 500 litres are used to produce 1 kg. of steel. In developed countries like USA, Japan and Germany only 60 litres of water is used for the same purpose. This is true for other commodities also.

Further it is also observed that in India the transmission loss is 30 to 40 per cent both in agriculture and civic bodies supply, while it is only 13 per cent in developed countries.

Sustainable Water Harvesting

Natural resources, viz., soil and water need to be conserved, developed and utilized efficiently due to ever-depleting ground waters the one feasible source of fresh water. The water tables are not getting recharged due to more runoff (35-45% of annual rain) during the monsoon season, which lasts only for three months (July-September). The major source of water available for either agriculture or human consumption is the precipitation that falls on the earth's surface. Since more than 80 per cent of the annual rainfall is received in a shorter span of three months, it becomes rather impossible to check this runoff water. Thus there is need to manage this running water so as to use it for irrigation and other routine purposes.

Water Harvesting

There is a vast potential to prevent rainwater to move down into the streams and rivers and ultimately into the sea. Efforts are continuously being made in this direction to conserve rainwater. Recently water harvesting, either in the impounding type or dug out reservoirs, is gaining importance. The water harvested during the rainy season can be used for one or two supplementary irrigations to the rabi (season) crops. Generally these reservoirs are made at the outlet end of the catchments drains out. The harvested water is then carried through underground pipelines to irrigate the command area down slope. This practice is important and a must, both for recharging the groundwater and for irrigation purposes. Even fisheries can be promoted in this harvested rainwater. However, the observations in the region have shown that these reservoirs get silted up within a few years of their construction. The reason being unabated soil erosion in the catchments area from where the detached soil particles are carried away by runoff water to the site of the harvested water where these particles settle down and ultimately silt up the whole reservoir. This result in damage to the agricultural land because of increased velocity of the runoff flow from the silted reservoirs.

Thus the process of water harvesting does not end with the simple construction of water reservoirs but the most important aspect of it is to prevent/check soil erosion in the catchment areas so as to achieve sustainable water harvesting. The depletion of groundwater will make our country more prone to earthquakes. Though it is not measures such a recharging of groundwater and the planting of trees should be undertaken. An appropriate mix of legislation pricing policy and enforcement measures is essential to optimize water conservation and protection.

REFERENCES

Journals (Quarterly)

Journal of Indian Water Works Association—Pipeline, Road, Vakola Santacruz (East), Mumbai-400 056.

Journal of Rural Development—NIRI, Rajendra Nagar, Hyderabad 500 038, India.

Water and Energy International—Central Board of Irrigation and Power, Mailcha man—New Delhi-110 021, India.

Waste Lands News—Society for Promotion of Wastelands Development, 40, Qutab Institutional Area, New Delhi-110 014.

22

WATERSHED MANAGEMENT APPROACH AND SUSTAINABILITY OF THE RENEWABLE NATURAL RESOURCES

(An Indian Experience and Lessons)

A. Abdulraheem and *A. Thaha Sahad*

Introduction

Sustainability of renewable natural resources in general and common pool natural resources (e.g., Land, Water and Forests) in particular has now become a major concern to natural resources policy-makers, planners, scholars and managers in both developed and developing countries of the world. The term "Sustainability" implies the abiiity of a natural resources system to produce socially optimum level of output, which is necessary to meet in perpetuity the needs and aspirations of the people dependent on the system with no detrimental effects on the resources system itself and the physical environment, and with no imposition of significantly greater risks on future generations. In other words, sustainability implies not only conserving natural products which are maintaining ecological functions and supply of natural resources products which are essential to the livelihoods of local people.

By a common property resource such as a village pasture, an open access resource such as marine fishery and sometimes even a private property resource such as fallow cropland. Although the CPRs include such diverse things as common pastures or grazing lands, community forests, community fish ponds, lakes, rivers, streams, groundwater basins, air sheds, etc., they all face one common problem and that is: how to co-ordinate the whole community. The occurrence to "the tragedy of the commons" implies loss of sustainability, which translates to loss of welfare on part of those who depend on the CPRs in question for their livelihood. Both developed and developing countries of the world are best with "the tragedy of the commons" and are in search of practicable strategies to resolve the problem.

In this paper, drawing upon India's experience, an attempt is made to show how the watershed management approach could attain sustainability of the renewable natural resources of land, water and forest in general and CPRs in particular.

Sustainablity as a Goal in CPR Management

Sustainability is being increasingly accepted globally as an explicit in the development and management of renewable natural resources in general and CPRs in particular. Sustainability of socially optimum level of output as a goal is superior to the goal of maximum sustainable yields (MSY) which, until recently, had been the major goal of management of many renewable natural resources. This is so because, unlike sustainability, the MSY principle does not consider ecological and other intangible effects of resource exploitation nor does it take into account the changing human needs over time.

Sustainability should, however, be made precept in natural resource management. Plato, for instance, accurately and graphically described direct and indirect effects of deforestation of the mountains of Attica on the region's soil and water resource and the economy of Athens[1].

Conditions for Sustainability

Dasgupta[2] mathematically derives two conditions for attaining sustainability. The first condition states that the rate of harvest from a renewable natural resource should not exceeded the rate of natural regeneration/replenishment at any point in time. In mathematical terms, this condition can be stated as : $Yt = H(St)$ where *Yt* denotes harvest rate of time, *t*, and H (St) is the rate of natural regeneration in time, *t*. This condition characterizes a stationary harvest policy.

The second condition states that the present value of marginal cost of increasing the current harvest. This condition is a perquisite for a stationary harvest policy to be optimal.

An optimal stationary policy which specifies an optimal level of stock, S^*, and an optimal rate of harvest, Y^*, has an underlying production function relating labour and capital inputs, natural resources stock, and technology to production of natural resources commodities. A typical natural resource commodity production function may be represented as follows[3]:

$$Y(t) = f[L(t), K(t), S(t), t]$$

Where $Y(t)$ is the natural resource commodity or the harvest/output of the natural resource at time, t, $L(t)$ and $K(t)$ are the labour and capital inputs respectively used in production of Y, $S(t)$ is natural resource stock and t indicates technology. The output of resource commodity could be increased by technological improvement of by augmenting the stock, or by intensifying labour and capital imputs. An improvement in technology will shift the production function upward and thereby may change the optimal stationary policy, ceteris paribus. Thus, a unique optimal stationary policy associated with every unique resource commodity production function may be visualized. Based on this relationship the optimal levels of harvest and stock need not be stationary or constant over time; they could change in response to human needs and aspirations. In other words, Natural resources systems could be manipulated and managed to produce socially optimum output over time on a sustainable basis.

Two approaches have been used to attain sustainability on the commons: the agro-ecological characterization approach and the watershed management approach. Both are quite similar in many respects.

The Agro-Ecological Characterisation Approach

Inter-disciplinary in ecology, climatology, geography, ecophysiology, soil sciences, plant science, animal science, environment economics and other resource related disciplines is needed to generate new information and technologies necessary for sustainability. At present there is no universally accepted conceptual framework to integrate these different disciplines so that they could interact effectively and contribute to the goal of sustainability. Agro-ecological characterisation provides a potentially practicable framework for effective integration of various disciplines.[4] Early global and regional attempts at agro-ecological zoning have shown that this approach can be valuable to agricultural planners. In essence, proper agro-ecological characterization depends upon the collection, organization and analysis of climate, soil and land topography data and their influences on the distribution of spices, plant growth, and agricultural yield. Despite significant advances made in the methodology of agro-ecological zoning, the application of this approach to effective management of natural resources is beset with a number of problems such as controversy over tools and techniques, lack of a unified approach across different disciplines, data inadequacies, and lack of trained staff[5]. In India, this approach is being adopted on a pilot project basis during into 15 agro-climatic zones and 74 sub-zones and agro-climatic zonal planning has been accepted as an instrument of ensuring comprehensive and integrated approach to natural resource management.

The Watershed Approach

This approach is conceptually very similar to the agro-ecological characterization approach. The only difference between the watershed approach and the agro-ecological approach is that in the former, watershed is used as a unit for planning and management whereas in the latter an agro-ecological zone serves this purpose. A watershed may be defined as a natural drainage area of a river, a tank, or a lake. A watershed approach, a watershed is used as a unit for planning and management of land, water, forests, and other resources of the watershed. The approach is holistic and multi-disciplinary and is a practicable approximation of the systems approach. It enables the planners and managers to consider simultaneously various physical, biological, socio-cultural, economic and institutional factors operating within a watershed and its surrounding environment and formulate a comprehensive and integrated watershed development plan to achieve specific private and social objectives. In the watershed approach, natural and human resources are all inter-dependent and interact with one another. A mono-disciplinary approach focusing on only one of the resources in isolation of the others is inadequate. Similarly, restoration and development of only privately owned land resources in an area cannot resolve the problem of degradation of the entire land resources in an area; CPRs of land which in a typical Indian dry land watershed account for nearly 30 per cent of the geographical area should also be restored and developed. This means that nothing short of a systems approach can realize the full potential synergistic benefits from the use of watershed's resources. The watershed approach is also justified on the ground that it internalizes various externalities.[6] Involved in the use of land water resources in a watershed and thereby narrows the hiatus between individual and social interest. The externalities can be internalized either by requiring the beneficiaries to compensate those whose activities produced the (positive) externalities or by disturbing the cost of watershed development among all farmers situated in the watershed in proportion to the size of their land holding.

A typical watershed development project consists of the following activities:

1. Assessing watershed dwellers felt needs, priorities, resources and constraints through a benchmark survey.
2. Survey, measurement, and mapping of the natural resources of the watershed, and assessment of their status and productivity.
3. Planning for restoration/development, conservation and optimum utilization of the watershed resources using the latest available technologies and in accordance with the watershed dwellers needs and preferences.
4. Provision of basic supporting infrastructure and creation of necessary institutions.

5. Planning for human resource development (through education, training and motivation), and utilization.

India's Experience and Lessons

In India, the watershed approach was first adopted on a significant scale in 1974 when the Government of India (GOI) enforced its implementation under a centrally-sponsored "Scheme of Soil Conservation in the Catchment's of River Valley Projects". These projects are now being implemented by the State Governments though their Agriculture Department and technical back up is provided by the All India Coordinated Research Project for Dry Land Agriculture (AICRPDA), the Central Research Institute for Dry Land Agriculture (CRIDA), and the Central Soil and Water Conservation Research and Training Institute. Another centrally sponsored scheme of Integrated Watershed Management in the catchments of Flood-prone Rivers was started during the sixth plan period. In July 1986, the Union Ministry of Agriculture and Rural Development launched the National Watershed Development Programme (NWDP) for rain-fed agriculture as a centrally sponsored scheme.

The Sukhomajri Experience

The Sukhomajri project is a well-known model of micro-watershed development in India and has been well documented. The Sukhomajri experience indicates that exhortations for participation and co-operation do not work, especially if they are aimed at people who live on the margin of subsistence[7]. The poor cannot stop grazing their animals in highly degraded and over-grazed common properly lands for the sake of land conservation when their lives depend on the animals. Only with increased productivity of crops and increased milk yields resulting from supplemental irrigation made possible by the reservoirs constructed under the project, and assurance of equal share of the reservoir water were the villages ready to invest in soil and water conservation measures and to participate in the programme whole-heartedly[8]. A major criticism of the project is that it is highly resource-intensive; a lot of technical, managerial, and financial resources have been spent on the project. Resources of that magnitude and quality are not available in India for replication of the model and hence it is no surprise that the model has not available in India for replication of the model and hence it is no surprise that the model has not been replicated elsewhere in India including its home state, Haryana.

The Ralegan Siddhi Experience

Like the Sukhomajri project, the Ralegan Siddhi project is another well-known and well-documented model of micro-watershed development in India[8]. The Ralegan Siddhi experience indicates that the rural people, under the guidance and leadership of good, enlightened, and honest persons, could achieve a lot through their own efforts and resource. Given proper leadership, it should be possible to replicate this model in other Indian Villages. Padma Shri Anna Hazare is now working to institutionalize the model by training rural youths who have volunteered to follow in his footsteps.

The Pidow Project Experience

The Karnataka's District Watershed Development Programme (DWDP) is also well known in India for its innovative, three-tier organizational structure, state-wide coverage, and laudable achievements. The PIDDOW project experience shows that a non-governmental organization could do a good job of enlisting people's participation in soil and water conservation programme perhaps better than a government agency[9]. As a first in the process of enlisting their participation, the project staff organized the people into small homogeneous groups/associations around income generating activities and motivated and trained the people to design, construct, repair and maintain various soil and water conservation structure on their own private land as well as on the common properly lands.

The Operational Research Projects (ORP) Experience

As mentioned earlier, 47 ORPs in integrated watershed development were launched in India during the Sixth Plan period. On the basis of these findings, they concluded that the watershed approach is the key for maximizing crop production on a sustained basis without any detrimental effects on the environment, that the ideal size of a watershed unit for efficient planning and management of resources is 400-500 ha, and that the involvement of farmers, technicians, development agencies, and administrations is essential in all stages of watershed development and management.

To sum up, the major elements of the watershed approach that help attain sustainability are as follows:

1. Restoration of degraded land resources through appropriate soil conservation and land reclamation measures.
2. Harvesting, storage, conservation, and optimal utilization of rainwater.
3. Use of land according to its physical suitability. This typically means that: (a) steep slopes and fragile land in upper researchers of watersheds, which are mostly CPRs, are used for growing trees, grassed and other permanent vegetation to produce enough biomass for the needs of the watershed community with respect to fuel wood, fodder, organic manure, etc., (b) relatively flat lands are used for production of foods crops, and (c) cash crops, and the lands in lower reaches of watersheds are used for storage of rain-water for supplemental/protective irrigation during the dry season.
4. Preparation of resource budgets, nutrient budgets and balancing of the budgets by recycling of renewable resources like biomass, solar energy, water, atmospheric nitrogen and other plant nutrients. This will ensure fulfilment of one of the conditions for sustainability, i.e., inputs and outputs must be balanced.
5. Control of pests and diseases by biological methods.

6. Processing of timber, minor forest produce and other biomass to add value to them and to generate employment opportunities for watershed dwellers.
7. Manpower planning and development of human resources through education, training, motivation and provision of information about new technologies and government policies, and programmes.
8. Determining optimal carrying capacity of watersheds in terms of human and animal population at the existing and prospective levels of technology and adjusting the existing population accordingly. This needs to be integrated vertically and horizontally with planning in other watersheds.
9. Organising watershed community along economic activities and motivating them to mobilize their resources, to manage their CPRs collectively, and establish institutions for equitable distribution of benefits from the CPRs and maintenance of the developed CPRs in good productive condition.

Conclusions and Implications

India now has the technical know-how, more important, the means to halt the processes that lead to degradation of natural CPRs. What appears to be lacking is the political will to apply the available technical knowledge, and a national policy and an appropriate institutional structure to plan, co-ordinate, implement, monitor, and evaluate watershed development programmes on a national scale.

Given the interdependencies among various natural and human resources and hence the existence of externalities in their use and management, the watershed approach which is a close approximation of the systems approach, is the most appropriate approach for planning and management of natural CPRs. Sub-watersheds measuring 400 -500 ha each could be used as units for planning and management but smaller administrative and/or socio-economic units should be used for implementation and monitoring purpose. The components of balanced resource budgeting, crop nutrients budgeting, and biomass recycling constituting the watershed approach largely ensure that the conditions for sustainability are fulfilled. Integrated development and use of privately-owned arable lands and CPRs of non-arable lands in a watershed can generate enough biomass for meeting the basic needs of local people and nutrients requirements of crops on a sustainable basis.

There is a need for a cross-sect oral and interdisciplinary to sustainability. In essence, it requires that all major actors and players in the development process, namely, policy-makers, hydrologists, soil scientists, agronomists, horticulturists, foresters, environmentalists, resource economists, sociologists and so on jointly devise watershed development and management strategies that are not only technically and economically viable but also socially and politically acceptable.

The involvement of the people in planning and implementation is essential for success of watershed development projects. To enlist people's participation, good local leadership, flexibility in project design and operational procedures, equity in the distribution of project benefits and cost sharing, support of a non-governmental and non-political organization and proper education and training of people are all essential. Furthermore, people would not generally participate in a project unless the expected private benefits from participation are markedly higher than the expected costs of participation.

NOTES

1. Colchester, Marcus (1994). "Sustaining Forests: The Community-based Approach in South and South-East Asia", *Development and Change*, 25(1).
2. Dasgupta, Partha (1982). *The Control of Resources*, Oxford University Press. Delhi.
3. Hardin, G. (1968). The Tragedy of the Commons, *Science*, 162: 1234-148.
4. Howe, Charles W. (1979). *Natural Resources Economist: Issues, Analysis and Policy*. John Wiley and Sons, New York.
5. Oakerson, R.J. (1986). "A Model for the Analysis of Common Property Problems," In National Research Council, Proceedings of the Conference on Common Property Resource Management, April 21-26, 1985, National Academy Press, Washington D.C.

6. Oram, Peter A. (1988). "Moving Toward Sustainability: Building the Agro-cological Framework," *Environment*, 30 (9).
7. Singh, Katar (1988). "Managing Dry Land Watershed Development Programmes: Lessons of Karnataka Experience." Research Paper #1, Institute of Rural Management, Anand.
8. Swaminathan, M.S. (1986). Foreword. In National Research Council, Proceedings of the Conference on common Property Resource Management, *op.cit.*
9. Wiebecke, Claus and Wiebke Peters (1984). "Forest Sustention as the Principle of Forestry—Idea and Reality, National Resources and Development, Vol.20. Institute for Scientific Co-operation, Tublingen.

23

BIO-ENERGY AND ITS IMPACT ON RICE PRODUCTION IN AGRICULTURE OF THOOTHUKUDI DISTRICT

D. Solomonraj

Agriculture is the primary sector of Indian economy and one of the major consumers of energy. The level of productivity of crops depends upon the energy input during various farm operations. There is a direct relationship between the energy use and agricultural production. Energy shortages and cost escalations will impede the growth of agriculture and its impact will be severed on crop production in the short run. In this context, efficient management of scarce energy resources can play an important role in maintaining the growth in agriculture. This calls for energy conservation with a view to minimize energy use without affecting the crop production.

In India, the shortage of commercial energy has created new problems in agricultural sector and it has posed a serious threat to our objective of producing 200 million tonnes of foodgrains to meet our future requirements. We have to bridge the demand-supply gap of energy for producing enough foodgrains, for the ever-increasing population. In the present paper, an attempt has been made to study the possibilities of using commercial sources of energy efficiently, the scope for using renewable sources in agriculture that may be eco-friendly and helpful in increasing the output and the possibilities of reducing energy consumption by adopting energy conservation techniques.

This paper considers the energy resource management in agriculture with special reference to rice production at selected villages of Srivaikuntam block of Tuticorin District in Tamil Nadu with the following objectives :

(i) To study the energy use pattern in rice production.
(ii) To analyze the influence of various energy sources on rice yield.
(iii) To find out the influence of bio-energy on rice production in the study area.
(iv) To analyze the specific problems of energy use in agriculture and to offer policy suggestions to improve the existing energy use pattern with a view to conserve energy sources.

Materials and Methods

A multi-stage random sampling procedure was adopted to select the respondents. Srivaikuntam was selected among the seven taluks in the district. Out of the six blocks in the taluk, Srivaikuntam block was taken for the present study. At the third stage, ten out of thirteen revenue villages were selected in the block.

Finally, the numbers of agricultural households of the selected villages were found out and depending on the number of households in each village, a random sample of size 97 was taken. The data were collected by personal interview through a pre-tested interview schedule.

Analytical Procedure

The sample households are stratified into different farm groups on the basis of the size of farm lands. There were two groups, viz., small farm (not more than 2.0 ha) and large farm (more than 2.0 ha). Simple statistical tools were used to study the energy use pattern across the different farm groups. The per year, per hectare energy use of commercial sources like seed, fuel, pesticides and fertilizers and non-commercial sources like farmyard manure, human power and animal power and further grouped as chemical energy, bio-energy, mechanical energy, animal energy and human energy were analyzed in energy term.

$$Y = a + b_1X_1 + b_2X_2 + b_3X_3 + b_4X_4 + b_5X_5 + b_6X_6 + u$$

Where,

Y = Output of paddy in kg.
X_1 = Mechanical energy in MJ
X_2 = Bio-energy in MJ
X_3 = Chemical energy in MJ
X_4 = Animal energy in MJ
X_5 = Human energy in MJ, and
X_6 = Farm size in hectare

'a' is constant, b_1 to b_6 are the respective regression coefficients to be estimated and 'u' is error term.

To make a better comparison of different sources of energy among farm groups, the energy units of various fuels were converted into mega joules (*MJ*) by using standard conversion rates (see Appendix-1).

Results and Discussions

In this section, a detailed analysis of energy consumption level in paddy is attempted. This section examines the energy flow in paddy and energy from paddy output, by various term categories among the different sources of energy.

Energy use Pattern in Rice Production

Energy flow in paddy production can be better explained by commercial and non-commercial sources of energy. Dhawan *et al.*, defined commercial energy as that which comprises those components of energy cake which are generally not produced at the farm level and the farmers have to pay cash to buy them. Non-commercial energy consists of those components of energy cake, which are mostly produced at the farm level and generally the farmers do not pay cash to use them. The commercial and non-commercial energy requirement for paddy cultivation is presented in Table 23.1.

Table 23.1 : Commercial and Non-commercial Energy used for Paddy-Farm size Groups (in MJ/Hectare)

Sl. No.	*Category of Farmers*	*Commercial*					*Non-Commercial*				
		Seed	*Mechanical*	*Pesticides*	*Fertilizer*	*Total*	*FYM*	*Human*	*Animal*	*Total**	*Total***
1.	Small	2119.84	3450.22	452.64	19849.53	25872.13	805.96	1715.22	496.98	3018.16	28,890.29
		(7.3	(11.9)	(1.6)	(68.7)	(89.6)	(2.8)	(5.9)	(1.7)	(10.4)	(100.0)
2.	Large	1573.57	1564.51	298.47	17654.94	21091.49	854.28	1556.71	250.14	2661.13	23,752.62
		(6.6)	(6.6)	(1.3)	(74.3)	(88.8)	(3.6)	(6.6)	(1.1)	(11.2)	(100.0)
3.	All	2024.10	3119.73	454.54	19464.91	25034.28	814.43	1687.44	453.71	2955.58	27,989.86
		(7.2)	(11.1)	(1.5)	(69.5)	(89.4)	(2.9)	(6.0)	(1.6)	(10.6)	(100.01

Note : Figures in Paranthesis indicate percentage to the total.

* Source-wise total

** Grand total

Source : Field Survey.

The paddy cultivation requires more commercial energy which contributes on an average 70 per cent of the total source-wise energy requirement in all the farm categories. The total source-wise energy requirement for paddy ranges from 28,890.29 MJ/ha. in large farm. Among the various energy inputs, fertilizer is found to be a highly required source and its share is 68.7 per cent by the small farm and 74.3 per cent by the large farm to the total source-wise energy consumption.

The relationship among yield, total energy, and commercial and non-commercial sources of energy for paddy is explained in Table 23.2. The table shows that the small farm consumes 28752.62 MJ/ha. with an average yield of 3510,10 kgs/ha, as against the large farm consumes 23,752,62 MJ/ha. with an average yield of 2,780.14 kgs/ha. on an average, the share of commercial energy is quite high (89.4 per cent) in comparison with non-commercial energy (10.6 per cent). Out of the commercial energy, chemical energy contributes 71 per cent whereas mechanical energy 11.1 per cent.

Influence of Different Energy Source on Productivity of Rice

The results of an analysis of variance of outputs of paddy showed that there was significant difference between small and large farm groups. Therefore, regressions were estimated for the two groups also. An examinations of zero-order correlations indicated presence of multi-collinearity in the groups. So, explanatory variables were selected for the two groups after avoiding the problem of multi-collinearity.

The coefficients of regression equation of animal energy which has a marginal influence of 0.04 kg where as chemical energy has a negative influence of - 0.01 kg. This implies that the increase in chemical energy would lead to a decline in the yield. The elasticity co-efficients of mechanical energy and animal energy show that one per cent increase in mechanical and animal energy would cause 0.44 and 0.21 per cent increase in the output produced (Table 23.3).

Table 23.2 : Relationship Among Yield, Total Energy and Commercial and Non-Commercial Sources of Energy for Paddy

Crops	*Farm Size*	*Yield (kg/ha)*	*Total Energy (MJ/ha) (CE+NCE)*	*Commercial Energy (MJ/ha) (M+C+S)*	*Non-Commercial Energy (MJ/ha) (FYM/HP/AP)*	*Chemical Energy (MJ/ha) (F+P)*	*Bio-Energy (MJ/ha) (S+FYM)*	*Mechanical Energy (MJ/ha) (Fuel+ele)*	*Animal Energy (MJ/ha) (AP)*	*Human Energy (MJ/ha) (HP)*
Paddy	Small	3550.10	28752.62	25872.13	3018.16	20302.07	2925.8	3450.22	496.98	1715.22
			(100.0)	(89.6)	(10.4)	(70.3)	(10.1)	(11.9)	(1.7)	(5.9)
	Large	2780.14	23752.62	21091.49	2661.13	17953.41	2427.85	1564.51	250.14	1556.71
			(100.0)	(88.8)	(11.2)	(75.6)	(10.2)	(96.6)	(1.1)	(6.6)
	All	3415.09	27989.86	25034.28	2955.58	19890.45	2838.53	3119.73	453.71	1687.44
			(100.0)	(89.4)	(10.6)	(71.0)	(10.1)	(11.1)	(1.6)	(6.0)

Note: Figures in Parentheses indicates percentages to the total.
CE = Commercial Energy, S = Seed, F = Fertilizer, P = Pesticides, M = Mechanical Energy, C = Chemical Energy,
NCE = Non-commercial Energy, Ele = Electricity, FYM = Farm Yard Manure, HP = Human Power, AP = Animal Power.
Source: Field Survey.

Table 23.3 : Regression Coefficients of Energy inputs on Output of Paddy–Farm Size Groups

Variables	*Category*	
	Small	*Large*
Mechanical Energy	0.01 19	0.4 374
	(0.0008)	(0.1715)
Bio-Energy	–	0.3196
		(0.2465)
Chemical Energy	-0.0106*	0.0279
	(0,0006)	(0.0166)
Animal Energy	0.0458	3.6635
	(0,2385)	(1.9316)
Farm Size	3504.2341**	350.7265
	(228.4882)	(383.7698)
Intercept	74.6609	-1155.8643
F-Value	96.1789**	10.4281**
R^2	0.8369	0.8258
SEE	658.1234	4033.2794
No. of Observations	80	17

Note: Figures in parenthesis represent the Standard Error.
** Significant at 1 per cent level
* Significant at 5 per cent level

The elasticity co-efficients of chemical energy is negative showing that an increase in the chemical energy by one per cent would reduce the production by 2.01 per cent. This analysis shows that chemical energy seems to be used excessively or improperly, output will not change if the chemical energy is reduced to a marginal extent. The variables bio-energy and human energy which have not been included in the model are multi-collinear with the variables land which has a significant and a high influence on output variation. This implies that bio-energy and human energy have considerable influence exerted through area.

The regression results for the large farm group showed that the coefficients of all energy inputs affected the production of paddy crop positively (Table 23.4). The coefficients of regression equation of animal energy has a significant influence of 3.66 kg on output variation. The variable human energy (X_5) which has not been included in the model is multi-collinear with the variable land (X_6) which has a significant and high influence on output variation. This implies that X_5 does have considerable influence exerted through area (X_6).

Table 23.4 : Elasticity Coefficients of Energy Inputs on Output Paddy Farm Size Groups

Variables	*Category*	
	Small	*Large*
Mechanical Energy	0.0009	1.0478
Bio-Energy	–	0.9749
Chemical Energy	-0.0423	0.6734
Animal Energy	0.0004	1.2738
Farm Size	0.7107	0.5677

Note: The elasticities are derived from regression co-efficients.

Problems and Prospects of Energy use in Agricultural Sector

The following are some of the possible solutions to the problems of energy use in agriculture:

(i) The actual use of chemical energy for cultivating one hectare under paddy is 17,301.52 MJ. But the standard use of chemical energy prescribed by the Agricultural Department is 9,889.59 MJ.
Now, it is clear that the farmers are using chemical energy exceedingly. They can very well save 7.411.93 MJ/ha. (43 per cent) the chemical energy without affecting the productivity of paddy.

(ii) The chemical fertilizers demand more fossil finds for their manufacture and moreover, continuous use of such chemicals without balancing therewith bio-fertilizers, in turn affects the soil quality and fertility. So, in order to preserve the soil quality and fertility and maximize the crop output, the farmers can increase the present use of 2506.71 MJ / ha. of bio-energy to 6199.2 MJ /ha. (147 per cent) as prescribed by the Agricultural Department in the case of paddy cultivation.

(iii) The improved agricultural implements such as ploughs, digging tools, seed-*cum*-fertilizer drills, and manually operated wheel goes (for weeding and hoeing) are expected to save 30-50 per cent of energy inputs. These implements can also reduce the time involved in various agricultural operations.

Conclusion

The results of the regression analysis show that bio-energy and animal energy have great potential to increase output of paddy in the study area. Moreover, the small farmers are using chemical energy exceedingly or improperly in paddy cultivation and output will not change even if the chemical energy is reduced to a marginal extent.

Policy Implication

1. It is observed from the study that bio-energy has great potential for increasing output, necessary steps are to be taken to induce farmers to use more of bio-energy which is also renewable, rather than chemical fertilizers, through extension education. This will help in achieving sustained growth of agriculture which is eco-friendly as well.
2. It is also observed from the study that farmers are using chemical fertilizers exceedingly or improperly in the production process. The farmers can be taught through extension education and framing regarding the standard use of fertilizers, manures and pesticides.
3. Agricultural Department can play a greater role in popularizing or marketing improved ploughs and digging tools among the farmers.

to the farmers at a subsidized price rather than free of cost with a view to conserve energy from over use.

REFERENCES

Annual Report (1990-91), "Energy Resource Assessment and Planning for a selected village in Tamil Nadu", Department of Bio-energy, College of Agriculture Engineering, Tamil Nadu Agriculture University, Coimbatore, pp. 1-35.

Dhanapal, D. (1991), "Energy use pattern in Agriculture in a changing Tribal Area—A Micro Level Analysis", *Urja*, Vol. 29, No.2, pp. 79-85.

Dhawan, K.C., Mittal, J.P. and Arora, B.S. (1992), "Role of Animal and Human Energy in the Village Eco-systems of India", *Economic Affairs*, Vol. 39, Qr. 1, pp. 31-33.

Dhawan, K.C. and Mittal, J.P. (1994), "Role of Fertilizers is Increasing at Yield of various crops in India", *Economical Affairs*, Vol. 37, Qr. 2, p. 95.

Doering, O.C. (1977), "An Energy based Analysis of Alternative Production Methods and Cropping System in the Corin Belt", *Report of the Agricultural Research Station*, Purdue University, West Lafyette, p.5.

Kombairaju, S. (1989), "Energy use Planning in Crop Production at Micro Level: A Programming Approach", (Ph.D. Thesis), Dept. of Agricultural Economics, Tamil Nadu Agricultural University, Coimbatore, pp. 70-76.

Project Report (1990), "Energy Resource Assessment and Planning of Villages Kaktia and Mashiara in West Bengal", Department of Agriculture Engineering, Indian Institute of Technology, Kharagpur, West Bengal, pp. 12-27.

Sing, A.J. and Steinbant, C.E. (1974), "Energy use in U.S. Food Systems", Science: 183: 307-316.

Shukla, D. *et al.* (1976), "Pattern of Energy use per hectare in different areas of District Bareilly, Uttar Pradesh", *Indian Journal of Agricultural Economics*, 31: 245.

APPENDIX-1

Energy Content of Fuels and Energy inputs

Items	*Unit*	*Energy Content (MJ/Unit)*	*Reference*
A. Fuels			
1. Electricity	kwh	11.93	1
2. Kerosene	1	37.57	1
3. Petrol	1	48.23	1
4. Diesel	1	56.31	1
5. L.P. Gas	kg	47.95	1
6. Firewood	kg	16.74	1

(Contd...)

(Contd...)

Items	Unit	Energy Content (MJ/Unit)	Reference
7. Dung Cake	kg	10.05	1
8. Agricultural Waste	kg	12.55	1
(i) Fertilizers:			
1. Nitrogen	kg	60.60	1
2. Phosphorus	kg	11.60	1
3. Potassium	kg	6.70	1
4. Farmyard Manure	Tons	300.00	1
(ii) Chemicals:			
1. Superior Chemicals	1/kg	120.00	1
2. Inferior Chemical:	kg	10.00	1
(iii) Labour:			
1. Male	h	1.96	1
2. Female	h	1.57	1
3. Bullock pair	h	10.10	1
(iv) Machinery (Fuel Use Only):			
1. Tractor	h	112.62	1
2. Sprayer	h	24.12	1
3. Oil Motor	h	48.23	1
4. Current motor	h	47.72	1
(v) Crop Outputs:			
Main Products:			
1. Paddy	kg	14.70	1
2. Banana	kg	11.80	2*
By-Products:			
1. Paddy Straw	kg	12.50	1
Seed:			
1. Paddy	kg	15.20	1
2. Banana	kg	7.10	2*

kwh - kilo watt hour, 1 - litre, kg - kilogram, h - hour.
1000 Joules - 1 MJ (Mega Joules)
1000 MJ = 1 GJ (Giga Joules)

* As energy conversion rates for banana fruits are not. available, the high food value of A 11.9 MJ/Kg. For fruits has been adapted to banana and also the energy content of photos seed of 7.1 MJ/kg to banana roots.

Sources :

1. Annual Report, "Energy Resource Assessment and Planning for a Selected Village in Tamil Nadu", Dept. of Bio-Energy, College of Agricultural Engineering, Tamil Nadu Agricultural University, Coimbatore, 1990 -91, pp. 1-35.
2. Project Report, " Energy Resource Assessment and Planning of Village Kaktia and Marshiara in West Bengal", Department of Agricultural Engineering India Institute of Technology, Kharagpur, West Bengal, 1990, pp 12-27.

24

WATER RESOURCE DEVELOPMENT IN INDIA—A PROSPECTIVE

Mrs. V. Rajam and *C. Parvathi*

Introduction

In spite of the surface of earth being covered mostly by water, over the past few decades water scarcity has emerged as a global problem. All life forms and human economic activities are critically dependent on water, the movement of which is governed by the global hydrological cycle. Humans have moderated such movements of water and made available large quantities of this resource at times and places to suit the needs of societies and meet the demand of economies. In course of time, the level of such human interventions has grown to such an extent quantitatively that the need for a more informed approach to the vital natural endowment is being recognized an articulated all over the world (Falkenmark *et al.* 1980; Wittifogel 1957).

At the global scale, "freshwater lakes and rivers, which are the main sources for water consumed by the human societies, contain on an average about 90,000 been cubic metres (BCM) of water, which is about 0.26 per cent of total global freshwater reserves" (Shiklomanov, 1993). The global withdrawal of fresh water has increased sevenfold between 1900 and 2000 (Gleick, 2000).

If this trend continues, the world may see a more than six-fold increase in the number of people living in conditions of water stress, from 470mn today to 3 billion in 2025 (Postel, 1999). With respect to its share of global water resources, India is regarded as a better endowed country, with a total annual precipitation of about 4,000bcm. This is about 4 per cent of the total average annual runoff in the world's rivers (NCIWRDP, 1999a). As reported by the NCIWRDP (1999a), India has annual available water resources of 1953 cut km. On the other hand, it is also true that, if the population of a country grows rapidly, there will naturally be a proportional reduction in the per capita availability of water. From this point of view, India is facing a regime of stress (Dyson, 2001), as the per capita availability of water declined from around 5177 cum in 1951 to 1869 cum in 2001. Given the projected rise in India's population by 2025, the per capita availability may drop further to below 1,000 cum. In terms of country-level indicators, such a situation would be considered as one of "water scarcity". This picture is not typical of India, but a rapid growth in population exacerbates the situation.

I
WATER RESOURCE OF INDIA

Surface waters and groundwater are parts of the same hydrological system. They are interlinked with and influence each other. This can be clearly seen from the graphical presentation of India's water resources (Fig. 24.1) (Chaturvedi, 1985).

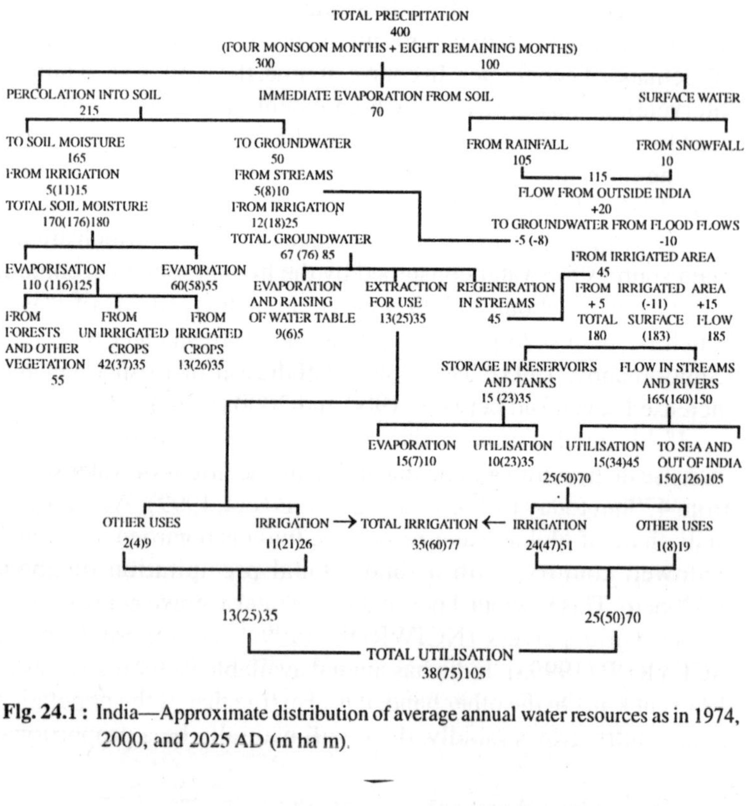

Fig. 24.1 : India—Approximate distribution of average annual water resources as in 1974, 2000, and 2025 AD (m ha m).

Notet : The numbers in brackets are those for the year 2000.
Source : National Commission on Agriculture, Chaturvedi, 1985.

It must also be said that our agricultural holdings are quite small and therefore large irrigation projects become quite expensive for this purpose, a Million wells scheme was started in 1988 so as to provide water to small and marginal farmers. The scheme did not work well in the initial years but in 1990, 20 per cent of the Jawahar Rozgar Yojana funds were specifically allocated for it. This allocation was raised to 30 per cent in 1993 to cover a whole lot of minor irrigation schemes as also the land development works. Ultimately, MWS was considered as an important programme for eradication of poverty in rural areas. The scheme is centrally sponsored and expenditure is shared between centre and the states on 80:20 basis.

Water Resource Scenario

In terms of the total annual precipitation per unit of land surface, India stands well above the global average. However, the figures of gross annual precipitation over the country as a whole provide an unrealistic picture of the actual situation with respect to availability of water in the country. The domination of the south-west monsoon in the making of the climate of South Asia results in a wide spatial variation in the levels of precipitation from the east to the west (See the Fig. 24.2) and an acute temporal variation through the concentration of heavy precipitation during the 2½ months of monsoon, spread over July, August and September. The variability is exemplified by a comparison of the number of rainy days in parts of Rajasthan along the north-western boundary, where it is just five, and in some areas in north-east where it rains for about 150 days (NCIWRDP 1999a:12).

Similarly, the average annual precipitation ranges from a low of about 200 mm in some locations in the western regions of India to a high about 11,000 mm in the north-east.

It is thus quite, clear that the averaged-out national picture of precipitation cannot provide any realistic picture of the actual problems of water resource development and management. Due to the spatial and temporal inequities in precipitation, rivers in most parts of the country show a skewed hydrograph. The rain rich regions experience regular annual inundations during the monsoon, while many rivers in the lower rainfall areas dry up soon after the monsoon is over. There is an increasing tendency to describe such obvious hydrological unevenness of rivers as disasters, whenever a river overflows its banks, statements accusing "too much rainfall" or "release of excessive water from the upstream" have become common. They are, however, not backed up by any open data. In areas with lower precipitation, growing water scarcity resulting from unsustainable utilization practices and vanishing conservation measures, are frequently blamed on "droughts". How much of the growing scarcity is the result of human intervention in the water systems and how much in due to metrological drought has not been openly analysed (Bandyopadhyay, 1989).

As humans increasingly started to intervene in the flow of rivers (Deursen, 2000), the gradual success of these interventions resulted in the upscaling of their operational dimensions that finally led to the emergence of large dams as instruments for river basin development (Bandyopadhyay *et al.* 2002). Like in all ancient civilizations, this scale of interventions in rivers and lakes has grown from very small to quite large over the last few centuries. In the colonial era, and more so in the post-colonial period, large scale supply side augmentation has been prescribed by engineers as the only way of meeting the growing water requirements of developing countries. As a result, policy-makers in India have been adhering to the idea of solving the water problem by harnessing the huge monsoon run-off in the main rivers in the eastern and north-eastern parts regions, primarily with the help of large structural interventions.

There is a huge gap between irrigation potential created and its actual utilization in respect of major and medium irrigation

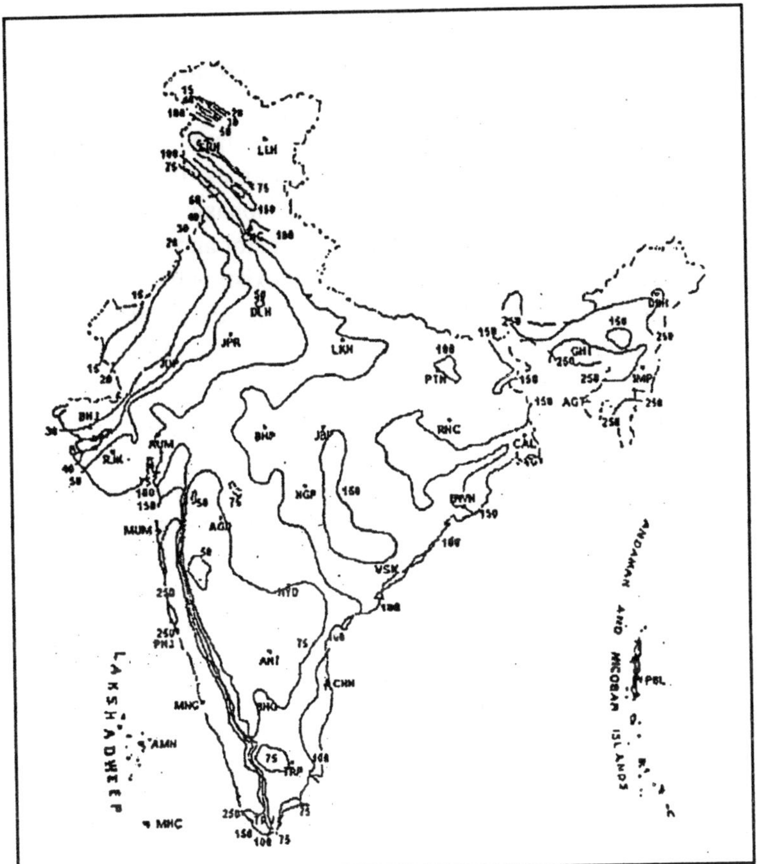

Fig. 24.1 : Spatial Variation in Precipitation Pattern
Source : http://www.imd.ernet.in/section/climate. animalrainfal.htm/

projects. This gap was 1.2 millions hectare in the First Plan period and it went up to 6 million hectares by the end of Eighth Plan period. As for the minor irrigation projects, such gap was of the order of 3.3 million hectares. Thus, the aggregate gap in the utilisation of the potential happens to be around 9.3 million hectares at present. This situation is one of the important reasons for shortfall in foodgrain production during the earlier years.

II
Optimum Utilisation of Irrigation

India's progress in agricultural production has largely been helped by the availability of irrigation water resources. It is, therefore, appropriate that the government has laid emphasis on completing the ongoing works in medium and major sector and on improving the quality of irrigation in the command of existing projects. Water as a resource going to be under tremendous pressure from population growth, rapid urbanization, industrialization and environmental degradation. A long-term perspective planning of water resources in thus the need of the hour to meet the various competing demands on water. The main objective of the national perspective plan for water resources, as the national water grid its new form is called, should be to gainfully utilize the surplus water available in some rivers by transferring it to river systems in the water-scarce tracts.

As is well known, our total water resources of 1150 cub km. per year is not uniformly distributed throughout the country. For example, 29 per cent of its available in the Brahmaputra basin which constitutes only 6 per cent of the country's area and where only 3 per cent of our population resides. As against this, the availability of water in much less than the national average in certain Southern and Western regions. In many Southern regions, for instance, the water availability is as low as one-fourth of the national average. The details of water resources potential in river basins of India is presented in Table 24.1.

III
Water Resource Availability and Requirement

India has about 16 per cent of the world's population but only 4 per cent of average annual runoff in the rivers of the planet. This compares with 8 per cent for China, which has about 25 per cent larger population than that of China. The utilizable annual water resource of India comes to 69m ha from surface and 43.2 m ha from groundwater resources, or a total of 112.2 m ha m per year.

Table 24.1 : Water Resources Potential in River Basins of India

Sl. No.	River Basin	Catchment Area Sq.km	Average Annual Potential m.ha m	Estimated Utilisable flow m.ha m	Estimated present use of surface water	Present stage of utiliza-tion percent
1.	Indus up to border	32128	7.33	4.60	4.0	87
2a.	Ganga	8,61,452	52.50	25.0	Data Classified	
b.	Brahmaputra	2,36,136*	58.56	2.4		
3.	Sabarmati	21,674	0.38	0.19	0.18	93
4.	Mahi	34,842	1.10	0.31	0.25	81
5.	Narmada	98,796	4.56	3.45	0.80	23
6.	Tapi	65,145	1.49	1.45	N.A.	
7.	Brahmni Baitarni	39,033	2.85	1.83	N.A.	
8.	Subernrekha	–	1.24	8.48	N.A.	
9.	Mahanadi	1,41,589	6.69	5.00	1.70	34
10.	Godavari	3,12,812	11.05	7.63	3.80	50
11.	Krishna	2,58,948	7.81	5.80	4.70	81
12.	Pennar	55,213	0.63	0.69	0.50	79
13.	Kaveri	81,155	2.14	1.90	1.80	95
14.	Other East flowing rivers between Mahanadi and Pennar		2.25	1.31	NA	
15.	East flowing rivers between Pennar and Kanyakumari		1.64	1.67	N.A	
16.	West flowing rivers of Kachch, Saurashtra		1.51	1.50	N.A.	
17.	West flowing rivers south of Tapi		20.09	3.62	N.A	
18.	Minor river basins draining into Bangladesh, Burma		186.93	69.03		

* Areas in Indian territory
N.A. Not available

Assuming the present population of the country to be 960 million, the per capita availability of utilizable water for the country as a whole is about 1170 m^3/capita/year. However, in Sabarmati, Kaveri, Pennar Sabarmati, and Kachch-Saurashtra basins it is much lower than average while it is higher in Brahmaputra, Barak, Ganga, Mahanadi, Brahmani, Baitarani and Narmada basins. Taking net cropped area as 144m ha the gross water depth available will be

0178 m per ha. Once again the average does not reflect the large disparities from basin to basin. The problems in water short basins are becoming increasingly serious, both in terms of domestic and industrial as well as agricultural requirements.

Projected Demand

The projected demand for the year 2000 and 2025 are as given in Table 24.2 (Hasan, 1996).

Table 24.2: Annual Water Demand in m ha m

Purpose/Year	*2000*	*2025*
Domestic use	3.3	5.2
Industrial	3.0	12.0
Energy	2.7	7.1
Irrigation	63.0	77.0
Others	3.0	3.7
Total	75.0	105.0

Source : Hydrological Chart.

By the year 2025, only 28 years away, it will be necessary to harness 100 per cent of the utilizable surface flow and 83 per cent of utilizable groundwater. Evidently a mammoth effort and determination are required to do so. It this is not done from now, there will be both food and water shortages resulting in serious social unrest.

Prospects for Improvements in Existing Systems

It has to be stressed that in a large federal country like India, all regions aim at balanced development. Water scarcity tends to put a limit not only on agricultural development but also on industrial development, urbanization and, overall economic growth. Hence, serious thought should be given to transfer of limited quantities of water to meet essential requirements of Pennar-Kaveri basins. This could be done at acceptable cost from Mahanadhi, and/or West flowing streams south of Tapi. Narmada development should take care of Sabarmati, Saurashtra and Kachch area.

IV
Interlinking of Rivers and Water Conflicts in South Asia

Water conflicts are frequently generated not by an inherent scarcity in a region, but over the sharing of additional supplies. The long-drawn Cauvery dispute in a case in poined. This is a conflict of interest between a downstream state (Tamil Nadu) that has a long history of irrigated agriculture and an upstream state (Karnataka) that was a late starter in irrigation development (but has been making rapid progress). The situation with respect to the canal in the north is no different. In all such cases, factors like end-use efficiency and sustainability of irrigation practices have given way to war over the quantity of water. In the present context of interlinking, however, it can be safely said that unscientific and wasteful uses of water cannot provide hydrological equity in a country.

Bandyopadhyay and Perween (2003) have argued that the proposed interlinking of rivers has the potential for generating four distinct types of conflicts. They are:

1. Compensation for resettlement and rehabilitation of the displaced,
2. Compensation for environmental damages from the project,
3. Sharing the benefits and costs of the project among the states, and
4. Cooperative management of the project in international river basins.

Conclusion

The proposed interlinking of rivers appears more to be a project for an *ad hoc* transfer of water looking for suitable justification. It cannot guarantee the security of domestic water supplies to the driver areas of India, in particular, the dry uplands, the only dependable solution to this problem lies in local-level harvesting and conservation of rainfall. If there is free competition, the proposed interlinking will not be able to compete with emerging technological solutions like desalinization, as far as domestic water supply in the coastal belt is concerned. The north of Bangladesh is already drying out after the Ganges was dammed by India in 1976. Now India is planning to do the same on (many of) the 53 other rivers that enter the country via India. Bangladesh depends completely on water ... We want no kind of war, but international law on sharing water is unsure and we would request the UN to frame a new law. It would be a last resort.

REFERENCES

Economic and Political Weekly, December 11, 2004.

Falkenmark, M.C Folue (2002): "The Ethics of Socio-Ecohydrological Catchment Management", *Hydrology and Earth System Sciences 6* (1): 1 - 9.

Food Demand Projections for India as quoted in NCIWRDP, 1999a: 51.

http: //www.telegraphindia.com

India's water and its Plausible Balance in Distant Future, NCIWPDP, (1999a: 29).

Integrated Water Resource Development: A plan for Action, National Commission on Integrated Water Resource Development Plan, MOWR, New Delhi.

Prabhu, Suresh P. (2003): "Interview" *In Indian Express*, New Delhi, March 2, (2004): Garland of Hope: River-Linking as a solution to water crisis" *Times of India*, Kolkata; August 14.

The World's Water 1998-1999 : The Biennial Report on Freshwater Resources, Island Press, Washington, D.

The World's Water 2000 - 2001: *The Biennial Report on Freshwater Resources,* Island Press, Washington, DC.

25

INTEGRATED WATERSHED DEVELOPMENT APPROACH TOWARDS PRODUCTIVITY AND INCOME GENERATION IN ERODE DISTRICT

S. Gurunathan, P. Namboothiripad and *P. Balaji*

Introduction

Land is the most important natural resource for meeting the nation's demand for food and for the growing food need of the fast (growing) exploding population to provide rural employment of large labour force, to meet the raw material needs of industrial sectors, to meet the requirement of cattle and to produce agricultural and horticultural crops to increase the income of the farmers and the nation.

Human and animal population is sailing but per capita productive capacity is decreasing due to soil erosion, desertification, salinization, water logging and improper management practices recurring droughts and floods which affect the soil fertility and land productivity. Per capita land availability in India has come down from 0.48 ha in 1952 to 0.15 ha in 2002. Higher investment for land resource management is, therefore, prime need of the hour.

Watershed Approach

Watershed is a geo-hydrological unit, which drains into a common drainage point. Watershed Development is a conscious effort to optimize and maintain the productivity of the land systems through integrated optimal development of water, land vegetation and livestock resources and appropriate land use practices, causing minimum environmental damage. Watershed development is an approach towards irrigation aiming at bringing back the harmony amongst the communities, rather than divided in the name of caste, creed, religion and class.

The technical high level committee headed by Dr. C.H. Hanumantha Rao (1994) appointed by the Ministry of Rural Development recommended the need to adopt watershed approach in all the area development projects so that development efforts could sharply be focused on a single eco-system and technical expertise can flow to a definite area with the local people participation in its management.

Objectives

The evaluation study was undertaken with the following objectives—

- To evaluate the impact of Watershed Development activities on productivity, income generation, and resource potential.
- To assess the effectiveness of the organizational arrangements for implementation.
- To find out the problems encountered and to collect the feedback information from the beneficiaries.

METHODOLOGY

Field study was taken up by interviewing the farmers from Gudhiyalathur watershed area of Erode district to study the impact of the programme and the collected information and data were tabulated for future analysis and interpretation.

The following procedure was adopted for carrying out this study—

1. Interview schedule to collect information from the beneficiaries and to evaluate the watershed development activities was prepared, protested and finalized.
2. A sample of 50 farmer beneficiaries from a total of 1406

beneficiaries were selected at randomly i.e., five beneficiaries from each watershed representing all the activities/treatments and interviewed with the well-structured interview schedule.

3. The data analyses were carried out by using average and percentage analyses besides using Garrett rating method.
4. Further conclusions were drawn and suggestions were made.

Schemes Implemented

The following activities are contemplated under watershed development—

- Soil and moisture conservation measures,
- Afforestation,
- Agro-forestry, and
- Distribution of fuel saving devices (improved *chulha's*).

Soil and Moisture Conservation Measures

Degradation of land due to surface runoff and erosion of soil is a major problem. During the north-east monsoon when water is available in abundance, the soil is unable to absorb it and the water goes waste as runoff eroding the valuable topsoil. Thus, in order to reclaim the degraded lands and make it fit for raising tree and annual crops, the following soil and moisture conservation measures are very essential :

(a) Gully Plugging (GP),
(b) Large Reek Check Dam (LRCD),
(c) Masonry Check Dam (MCD),
(d) Runoff Management (ROM),
(e) Percolation Pond (PP), and
(f) Contour Bunding.

These measures are intended to benefit the entire area by preserving the top soil and thereby increasing capability of lands to sustain vegetative growth as well as increasing ground and surface water availability thereby benefiting the people in and around the area.

Soil and Moisture Conservation Works

Tractor ploughing was taken up in the project area in four sub-watersheds only. This activity was taken up to benefit 221 beneficiaries. The average area benefited by a farmer worked out to 2.25 ha with a cost of Rs.612.71 while the per hectare cost being Rs.272.25 (Table 25.1). Farmers are ready to contribute even more to take up this work in an extended area.

Table 25.1: Tractor Ploughing

Particulars	*Tractor Ploughing*
Total number of beneficiaries (Nos)	221
Watersheds covered (Nos)	4
Total area covered (Ha)	497.37
Total cost of treatment (Rs.)	1,35.41
Area covered/beneficiary (Ha)	2.25
Cost/beneficiary (Rs.)	612.71
Cost/ha (Rs.)	272.25

Source : Compiled from the particulars finished by Agri. Engg. Dept., Satyamangalam.

Contour Bunding

Contour bunding is essential in order to conserve the long-term productivity of the land. Contour bunding occupies important place in the different types of treatments to conserve soil and water resources. Contour bunding has been done in seven sub-watersheds out of ten. The area treated in each sub-watershed varied from 100 to 337 ha. The total cost utilized for this treatment was Rs.14,12,791 and the unit cost varies from Rs.1,126 to 1,372 (Table 25.2).

Table 25.2 : Details of Contour binding over various watersheds

	Contour bunding		
Name of the Watershed	*Area (ha)*	*Cost (Rs.)*	*Unit Cost (Rs./ha)*
Ekkathur	100.720	123767	1229
Pudupalayam	42.045	57675	1372
Irutipalayam	320.94	423372	1319
Periyasalatti	156.190	175944	1126
Kadambur	194.895	230049	1180
Pavolakkuttai	337.765	401984	1190
Kadaganalli	294.330	–	–
Total	446.885	1412791	7416

Source: Compiled from the particulars furnished by Agri Engg. Dept., Satyamangalam.

Drainage-line Treatment

Major Check-dams and Run-off Management

The Agricultural Engineering Department has constructed a total of 18 check-dams, 11 Run-off management activities and one percolation pond. The check-dams and ROM structures have been constructed based on the availability and position of the land area in the sub-watersheds (Table 25.3).

Agro-forestry

Under Agro-forestry, it is proposed to distribute fruit/miscellaneous tree seedlings for creating plantations on private lands. Agro-forestry aims at raising tree plantation to meet the fuel wood and small timber requirement of the community and also the fodder needs of the livestock.

In the project area, afforestation works are to be carried out both in the degraded hillock area as well as in waterlogged tank beds etc. The villagers are allowed to remove lops and tops from

Table 25.3 : Watershed-wise details of Major Check-dams

Name of the Sub-watershed	*Check-dam*	*ROM*	*PP*
Ekkathur	1	3	–
Basuvanapuram	6	3	–
Pudupalayam	1	–	–
Irutipalayam	–	1	–
Kadambur	5	1	–
Periyasalatti	2	1	–
Pavolakkuttai	3	1	–
Kadaganalli	–	–	1
Kadathur	–	1	–
Total	18	11	1

Source: Compiled from the particulars furnished by Agri. Engg. Dept., Satyamangalam.

the trees. Further, when the plantations reach maturity, it will be auctioned in public and the sale proceeds shared between the department and the local body in the ratio of 40 : 60.

People Participation

People's participation is one of the crucial determinants for the sustanability of any kind of developmental programme. The success of watershed management depends largely on effective people's participation at various stages of implementation.

Formation of village level committees in the village where watershed activities were implemented ensures people's participation and their involvement from planning stage itself. The Self Help Groups formed by one NGO in the project area was functioning as village level committees. But the SHG's formed were not able to continue any work due to the non-cooperation within the members, problems related to NGO etc.

In this connection, several meetings and campaigns were conducted by various line departments to create awareness and involvement in the project on improved technologies. From the information collected from the line department officials and user group members during the field study, it was observed that the farmers who were initially reluctant were showing positive signs of participation and involvement and are keen to accept the developmental activities. The farmers whole-heartedly welcomed the soil and water conservation works, agro-forestry and horticulture plantation especially.

No action was taken in this watershed to form the Watershed Association, Watershed Committee and Watershed Development Team. The four line departments were taken their own risk and made all arrangements in the people participation and developmental activities implemented.

IMPACT EVALUATION OF THE WATERSHED DEVELOPMENT ACTIVITIES

The Government of India guidelines for watershed development have set-forth certain criteria for measuring the success of watershed development projects in terms of technical and general aspects. Impact-evaluation of Gudhiyalathur Watershed in Sathyamangalam block of Erode district was taken up in terns of technical and general categories. The evaluation was done through a set of bio-physical, socio-economic and participatory indicators.

Evaluation of People's Participation in Watershed Development Programme

People's participation plays important role for the success of any developmental project. Especially in wasteland development activities, the role of people is very much essential, because the real benefits will go to the farmers. Necessary efforts have been made in this study to get the feedback about the various watershed development activities carried out under this project. The evaluation study include farmers opinion about functioning of user groups, accessibility to DRDA staffs, extend of technical support from DRDA, exposure visits and their usefulness and also the overall impact of various watershed activities on agricultural production. The results of the feedback analysis are given below.

Awareness of People

Awareness of the people about the watershed development activities is evaluated and presented in Table 25.4. The study revealed that most of the people were aware of the watershed activities implemented in their watershed area. Among the respondents more than 82 per cent got information from the DRDA officials. About 19 per cent got their information through friends and relatives.

Table 25.4 : Source of Information about Watershed Activities

Source of Information about Watershed Management	*DRDA officials*	*Friends/Relatives*	*Total*
Respondents (Nos)	41	9	50
Per cent	82	18	100

Though the farmers have generally become aware of the scheme, still more efforts are to be required to make them fully aware of their roles and responsibilities and to have better involvement in execution and maintenance work of the programme. From the above results, it is clearly observed that the line departments have done good efforts to make the farmers much aware of the watershed programme.

Exposure Visits and their Usefulness

Two exposure visits each comprised 100 User Group (UG) members were taken to Coimbatore and Dindigul. During this visit, the UG members were exposed to various watershed development and treatment activities. Out of the 50 respondents only 17 (34.0 per cent) attended the exposure visits and gained knowledge.

Usefulness of the exposure visit is presented in Table 25.5.

Exposure visits help in getting awareness and knowledge regarding new technical know-how and benefits of various watershed treatment activities among the members. During the survey, it is observed that the farmers need more exposure and field visits and it will help the farmers to further maintenance of the watershed activities.

Table 25.5: People's Opinion about the Exposure Visits and their Usefulness

Particulars	*No. of Respondents*	*Per cent*
Attended Exposure Visit	17	34.0
Not Attended Exposure Visit	33	66.0
Useful	17	100.0
Not useful	0	00.0

Participation in Decision-making and Implementation

Participation of farmers in decision-making and implementation of the activities is presented in Table 25.6. Farmers of Gudhiyalathur watershed have actively participated in the implementation of the watershed development activities such as construction of percolation pond, check dams, gully control structures and plantation activities. Members of the UGs along with the DRDA staffs are involved in the decision making process.

Table 25.6 : Participation of UGs in Decision-making and Implementation

Particulars	*Members of User Groups (in Nos)*		*Total*
	Participated	*Not-participated*	
In Decision-making	32 (64.00)	18 (36.00)	50 (100.00)
In Implementation	48 (96.00)	2 (4.00)	50 (100.00)

Notes : Figures in parentheses indicate percentage to the total.

Functioning of User Group

Opinion of the respondents about the functioning of the user group is presented in Table 25.7. Among the respondents 84.0 per cent reported that functioning of the user group is good and about 8.0 per cent expressed very good. Fifty-six per cent of the user group members expressed that there is good interaction among the members. Cent per cent of the respondents opined that the usergroup functioned good in site selection and the nature of treatment to be carried out.

Table 25.7: Respondents View about the Activities of the User-group

Opinion	*Very good*	*Good*	*Average*	*Poor*	*Total*
Functioning of UG	4 (8.0)	42 (84.0)	4 (8.0)	–	50 (100.0)
Members Interaction with other Members	6 (12.0)	28 (56.0)	16 (32.0)	–	50 (100.0)
Site selection for activities/treatment	–	50 (100.0)	–	–	50 (100.0)
Nature of work to be carried out	–	50 (100.0)	–	–	50 (100.0)
General Participation in UG meeting	–	40 (80.0)	10 (20.0)	–	50 (100.0)
Participation in Group Activity	4 (8.0)	34 (68.0)	12 (24.0)	–	50 (100.0)
Interaction of UG with PIA Officials	–	45 (90.0)	5 (10.0)	–	50 (100.0)
Participation in Common Activity	–	38 (76.0)	12 (24.0)	–	50 (100.0)

Note : Figures in parentheses indicate percentage to the total.

With regard to the participation of people in user group meeting, 80.0 per cent of the respondents and 20.0 per cent told that participation of people was only average. Sixty-eight per cent of the people in the user-group opined that the participation in group activity was good. This is followed by average (24.0 per cent) and very good (8.0 per cent).

Among the respondents, 90.0 per cent of the usergroup members reported that there was good interaction between UG and DRDA officials and only 10.0 per cent expressed that, there is only average interaction. People's participation in common activity is good, as reported by 76.0 per cent of the respondents.

OVERALL IMPACT OF WATERSHED DEVELOPMENT ACTIVITIES

Impact on Well-water Recharge

Construction of water conservation measures such as check dams, percolation pond and runoff management structure have enhanced the storage capacity in the watershed to store the excess runoff for surface water storage and groundwater recharge. It indicates that a total of 3000 ha of land have been benefited which shows that the watershed development programme has helped in improving groundwater recharge, water availability for livestock other non-domestic uses and prevent the soil erosion in this watershed.

Impact of watershed development activities on the water level in the wells and ponds are presented in Table 25.8. The data revealed that the water level in the wells has increased from minimum level of 1.5 feet to 2.0 feet in September 2003. The data indicates that the overall water level in the wells has increased after construction of percolation pond as compared to the period before scheme implemented.

Table 25.8 : Impact of the Programme on Well-water Recharge

Year	*Months*	*Average Rainfall (mm) in Sathy Block*	*Water-level in the open Wells (ft)*	*Water-level in the Ponds (ft)*	*Recharge-level in the Wells (ft)*
2002	October	158.0	5.2	4.5	5.2
	November	39.0	3.0	3.0	3.0
	December	8.0	1.5	0.0	1.5
2003	January	0.0	1.5	0.0	1.5
	February	8.0	1.0	0.0	1.0
	March	146.4	3.1	2.7	3.1
	April	12.0	3.8	1.5	2.8
	May	18.8	2.6	1.0	2.6
	June	26.0	2.6	1.0	2.6
	July	14.0	2.3	0.0	2.3
	August	101.0	3.2	2.3	3.2
	September	3.0	2.0	1.2	2.0

Farmers' opinion with regard to the water conservation measures clearly indicates that there was increase in the water level in the wells due to the construction of check-dams, percolation ponds and run-off management structures.

Cropping Pattern and Cropping Intensity

The change in the cropping pattern and cropping intensity after the implementation of watershed development activities are presented in Table 25.9. In addition to the crops like ragi, maize, samai, sorghum and gingelly, the farmers cultivate the new crops like tapioca, sugarcane, groundnut and fruit trees like mango, coconut, kapak and tamarind. The farmers also tried the cultivation of marigold.

Table 25.9 : Impact of Watershed Development Activities on Cropping Pattern and Cropping Intensity.

Cropping Patten		*Cropping Intensity (%)*	
Before	*After*	*Before*	*After*
Ragi, Maize, Samai, Coconut, Sorghum, Gingelly	Ragi, Maize, Samai, Tapioca, Ground nut, Mango, Tamarind, Amla, Coconut, Kapak, Sugar cane, Marigold	100	103.01

Income of the Farm Households

The sources of income for the farm households are from crop, livestock, trees, off-farm and non-farm. There was considerable increase in the income of the household from crop and trees only. Of-farm and non-farm income remain constant and there was no decrease or increase. In general there was a little increase occurred in the income of the households, after the implementation of the programme. The plantation done in the programme, will give the appreciable yield in another two to three years only. The increase in the income from crop and trees was 80.81 and 89.85 per cent respectively (Table 25.10).

Table 25.10 : Impact of Watershed Development Activities on Income of Farm Households

Source of Income	*Before*	*After*
Crop income	26118	47226
Livestock	1128	1128
Trees	4328	8217
Of-farm	32750	32750
Non-farm	15800	15800
Household income	80124	105121

Productivity of Major Crops

Productivity of major crops and their yield difference per ha is presented in Table 25.11. There is a little increase observed in cropping intensity by 3.01 per cent. In addition to this there is 29.54 per cent increase in the productivity of ragi and 23.09 per cent increase in the productivity of maize. The productivity of the new crops like tapioca and groundnut is also grown with good yield.

Table 25.11 : Productivity of Major Crops and their Yield Difference (kg/ha)

Crop	*Before*	*After*	*Percentage-change*
Ragi	1100	1425	29.54
Maize	4200	5170	23.09
Tapioca	–	17348	–
Groundnut	–	976	

Problems Faced by the Beneficiaries

The problems faced by the beneficiaries during the project implementation were enquired and presented in Table 25.12. The result revealed that there was none of the problems faced by the beneficiaries from the project implementing agency staff or among the user-group members. Very less percentage of beneficiaries opined that they were faced problems due to lack of awareness about the programme.

Table 25.12 : Problems Faced by the Beneficiaries during the Project Implementation

Particulars	*Average*	*Low*	*Very low*	*Total*
Problems from PIA	–	7 (14.0)	43 (86.0)	50 (100.0)
Problem from UG members	–	3 (6.0)	47 (94.0)	50 (100.0)
Problem faced due to lack of awareness about the programme	4 (8.0)	–	46 (92.0)	50 (100.0)
Extent of problem faced by responsibilities due to delay and receiving funds	–	5 (10.0)	45 (90.0)	50 (100.0)

Note : Figures in parentheses indicate percentage to the total.

Benefits Obtained after Implementation

With regard to soil fertility, about 20.0 per cent of the respondents opined that there was very good improvement and 76.0 per cent as good. 24.0 per cent of the respondents told that there was very good improvement in the groundwater and 78.0 per cent responded as good in respect of employment opportunity (Table 25.13).

More than half of the respondents opined as good about the fodder improvement and three-fourth of the respondents expressed

Table 25.13 : Benefits Obtained After Implementation

Particulars	*Very good*	*Good*	*Average*	*Total*
Soil fertility	10 (20.0)	38 (76.0)	2 (4.0)	50 (100.0)
In bio-mass addition	–	38 (76.0)	12 (24.0)	50 (100.0)
Fodder improvement	–	28 (56.0)	22 (44.0)	50 (100.0)
Groundwater	12 (24.0)	38 (76.0)	–	50 (100.0)
On employment	11 (22.0)	39 (78.0)	–	50 (100.0)
On their income	–	40 (80.0)	10 (20.0)	50 (100.0)

as good in case of biomass addition. The study revealed that there was additional income of Rs.21,108 per ha generated from crops and Rs.3,889 from trees.

Opinion about Overall Impact of Watershed Activities

The overall impact was assessed by the opinion of the beneficiaries about the impact of watershed development activities. The study results revealed that the watershed development activities were very much oriented on the soil and water conservation followed by groundwater recharge and helps to alter the cropping pattern.

The order of the impact is as follows :

1. Soil and water conservation
2. Improved soil fertility
3. Cropping pattern
4. Increase in cropping intensity
5. Resource conservation
6. Yield increase
7. Groundwater recharge
8. Farm diversification
9. Fodder availability
10. Facilitates in organic farming
11. Increased fuelwood supply

Garrett Analysis

To study the level of impact of the watershed development activities, Garrett analysis was done and scoring was given to various benefits obtained from the programme (Table 25.14). The results revealed that among the benefits the highest score was recorded to the soil and water conservation improved. This in turn indicated that the construction of different water conservation measures such as percolation ponds, check-dams, Runoff management structures etc., have much impact on the conservation of water and control soil erosion. Soil fertility received next score. As indicated by the score

Table 25.14 : Garrett Score Value for the Benefits Obtained from Different Watershed Activities

Sl.No.	*Benefits*	*Score Value*
1.	Soil and water conservation improved	76
2.	Soil fertility improved	69
3.	Fuelwood supply	67
4.	Altering the cropping pattern	59
5.	Cropping intensity	56
6.	Yield increase	47
7.	Groundwater recharge	43
8.	Resource saving	33
9.	Fodder availability	26

values implementation of the watershed development programme has significant influence on the fuelwood supply, altering the cropping pattern and cropping intensity also.

Conclusions

Summary of Findings

Impact evaluation study on watershed activities was conducted for the Gudhiyalathur watershed, Sathyamangalam block of Erode District. The conclusions and suggestions of the study are presented below :

- As against the physical target of 4000 ha, the area covered and implemented the watershed activities up to nearly full target that is 3935 ha. This shows that the appreciable efforts of implementing agencies.
- Soil and moisture conservation works regulated the beneficial effects on recharge of groundwater, control soil erosion etc. In addition to this the cropping pattern and cropping intensity was changed with good effects and productivity of the major crops also increased.

- Exposure visits were found to be useful and the farmers needed additional visits to get more exposure.
- Even though the peoples participation occurred during the implementation of the programme, the community organizational activities like Watershed Committee, Watershed Development Team, Self-Help Groups, etc. were not formed in this programme.

Suggestions

- Watershed Committee, and self help groups should be formed to maintain and harvest the full benefits to be generated by the watershed development activities.
- Due to the highly sloppy nature of the project area the run-off will be more, hence additional percolation ponds may be constructed will be highly useful to the farmers.

APPENDIX

Watersheds Details

Sl. No.	*Name of the Sub-watershed*	*Total Extent (Ha)*	*Area Treated (Ha)*	*No. of Beneficiaries*
1.	Kadaganalli	733.95	460.16	150
2.	Karalayam	1378.42	356.10	102
3.	Ekkathur	742.51	455.57	162
4.	Basuvanapuram	526.81	460.35	168
5.	Kanakunthur	692.40	389.60	151
6.	Pudupalayam	748.01	228.01	81
7.	Iruttipalayam	1146.18	418.00	139
8.	Pavalakkuttai	1067.38	600.87	160
9.	Kadambur	961.92	407.06	145
10.	Periva Salatti	629.18	402.18	148
Total		8629.76	4179.9	1406

Source : Particulars furnished by ADA Office, Sathyamangalam.

Index